Praise for ANIMAL WISE

"*Animal Wise* is a thought-provoking and highly engaging set of essays that captures the changing views of scientists toward the minds and emotional lives of animals. It is sure to have broad impact on attitudes toward other species and our treatment of them. Thank you, Virginia Morell, for adding legitimacy to what we have so painstakingly observed."

—Joyce Poole, PhD, codirector of ElephantVoices, member of the Amboseli Elephant Research Project, and author of *Coming of Age with Elephants*

"In sprightly and clear prose Virginia Morell enters the world of animals with respect and insight and with the compelling argument that our lives differ only in degree. The recognition that we are bound in mind to many other creatures, all of them dependent on us for survival, will, I hope, arouse our compassion and assure them a future. This is a fascinating, timely, and important book."

—George B. Schaller, Panthera and the Wildlife Conservation Society

"From ants to apes, *Animal Wise* covers wide-ranging scientific research on the cognitive and emotional capacities of many different nonhuman animals. Noted author Virginia Morell writes clearly and concisely, and this easy read will surely be good for animals because we must use what we know about them to make their lives better in an increasingly human dominated world."

—Marc Bekoff, author of *The Emotional Lives of Animals* and *The Animal Manifesto* and editor of *Ignoring Nature No More*

"It is nice to see a science writer of Virginia Morell's distinction take on this increasingly important topic, and it is good to have her calm and careful voice added to the conversation. She has a great deal to teach us about the latest research on the frontiers of this fascinating new world. *Animal Wise* is a fine book."

—Jeffrey Masson, author of *When Elephants Weep*

ANIMAL WISE

ALSO BY VIRGINIA MORELL

Ancestral Passions: The Leakey Family and the Quest for Humankind's Beginnings

Blue Nile: Ethiopia's River of Magic and Mystery

Wildlife Wars: My Fight to Save Africa's Natural Treasures (with Richard Leakey)

ANIMAL WISE

THE THOUGHTS AND EMOTIONS OF OUR FELLOW CREATURES

VIRGINIA MORELL

CROWN
New York

Published in the United States by Crown Publishers, an imprint of the Crown Publishing Group, a division of Random House, Inc., New York.
www.crownpublishing.com

CROWN and the Crown colophon are registered trademarks of Random House, Inc.

Some of the material in this work was adapted from the following: "Minds of Their Own" in *National Geographic* (March 2008); "Going to the Dogs" in *Science* (August 2009); "Watching as Ants Go Marching—and Deciding—One by One; Profile of Nigel Franks" in *Science* (March 2009); "Why Do Parrots Talk? Venezuelan Site Offers Clues" in *Science* (July 2011); and "Inside the Minds of Cats and Dogs" in *National Geographic Special Edition: Cats and Dogs* (Spring 2012).

Library of Congress Cataloging-in-Publication Data
Morell, Virginia.
Animal wise : the thoughts and emotions of our fellow creatures / Virginia Morell.—First edition.
pages ; cm
1. Cognition in animals. 2. Human-animal communication. I. Title.

QL.785.M655 2013
591.5'13—dc23

2012031503

ISBN 978-0-307-46144-5
eISBN 978-0-307-46146-9

Printed in the United States of America

Illustrations on p. iii and p. 206 by Maria Elias
Jacket design by Christopher Brand
Jacket photography: John Lund

10 9 8 7 6 5 4 3 2 1

First Edition

For my Mother, and for Michael,

who loves dogs, cats,

and all the wild creatures.

And for our pets,

Buck and Nini,

who stayed close while I wrote.

Surely, the most important part of an animal is its *anima*, its vital spirit, on which is based its character and all the peculiarities by which it most concerns us. Yet most scientific books which treat of animals leave this out altogether, and what they describe are as it were phenomena of dead matter.

HENRY DAVID THOREAU

contents

introduction

Our organ of thought may be superior, and we may play it better, but it is surely vain to believe that other possessors of similar instruments leave them quite untouched.

STEPHEN WALKER

Animals have minds. They have brains, and use them, as we do: for experiencing the world, for thinking and feeling, and for solving the problems of life every creature faces. Like us, they have personalities, moods, and emotions; they laugh and they play. Some show grief and empathy and are self-aware and very likely conscious of their actions and intents.

Not so long ago, I would have hedged these statements, because the prevailing notion held that animals are more like zombies or robotic machines, capable of responding with only simple, reflexive behaviors. And indeed there are still researchers who insist that animals are moving through life like the half dead, but they're so . . . 1950s. They've been left behind as a flood of new research from biologists, animal behaviorists, evolutionary and ecological biologists, comparative psychologists, cognitive ethologists, and neuroscientists sweeps away old ideas that block the exploration of animal minds. The question now is not "Do animals think?" It's "How and what do they think?"

Hardly a week goes by that doesn't see a study announcing a new discovery about animal minds: "Whales Have Accents and Regional Dialects," "Fish Use Tools," "Squirrels Adopt Orphans," "Honeybees Make Plans,"

"Sheep Don't Forget a Face," "Rats Feel Each Other's Pain," "Elephants See Themselves in Mirrors," "Crows Able to Invent Tools," and (for me, as a dog lover, a favorite) "Dog Has Vocabulary of 1,022 Words."

How do scientists know that a dog has such an impressive vocabulary, that moths remember they were once caterpillars, that blue jays regard other jays as thieves, or that not only whales but cows, too, have regional accents? How can we prove that animals think? Once we have done so, what does that tell us about our relationships with them, and what does it tell us about ourselves?

Many of us have had some experience—playing with a pet or watching wildlife—that made us think an animal was planning something, or feeling joyful or sad. My husband and I are sure that our dog smiles, especially when he's playing with us, or when we give him a promised bone, or when we all are reunited after one of our business trips. We laugh with delight to see his joy and say things like "Look at how happy Buck is. He's really excited about getting his treat." But is he? Without language, is there any way to rule out what else he might be thinking about? Maybe he just caught a whiff of a squirrel's scent, or some of our old socks, or maybe he's not smiling at all but simply doing his best to imitate an expression he often sees on our faces and perhaps associates with bones or walks but doesn't fathom in the slightest.

Although, like many pet owners, I've often had the gut feeling that my dog and cats have mental and emotional lives, I've never tried to prove this—I'm a science writer, not a scientist. The only "proof" I can offer is that my pets' behaviors, activities, and facial expressions all suggest thoughts and emotions. Isn't that one reason, perhaps the main reason, we have pets in our homes? We want the company of lively, expressive creatures, beings that can be fun and loving, grumpy and bored, and that relate to and respond to us as only another living being can. In short, we want to be around something more than pet rocks.

I don't think I've ever met a pet owner who didn't have a story about his or her smart dog or clever cat. Probably we like having smart pets because, like smart people, they're interesting and entertaining. And sometimes our smart pets make *us* think.

In many ways, it was because of our first very smart dog, Quincie, a mixed-breed collie, that I began thinking about writing this book. When she was a puppy, Quincie liked to carry a pine cone in her mouth on our daily mountain hikes. I don't know why she enjoyed this, but at the trailhead she always searched among the cones and picked out one to take with her. One day, as we hiked up a steep path, she suddenly stopped, set her cone down, and nudged it over the side of the trail with her nose. She watched intently as the pine cone tumbled down the slope, and when it reached a certain momentum raced after it as if she were chasing a rabbit. She had imagined a game, *invented* it, and she played it almost every time we hiked that trail.

"She has an imagination!" I remember saying to my husband the first time Quincie did this. I was surprised, even though, of course, she also played imaginary games with us, as most dogs do, barking and pretending to be a "mean" dog when we chased her—though all the while she was also wagging her tail and giving us other signs that this was just for fun. My cats, too, delight in chasing balls, fabric mice, feathers or bits of cardboard on a string—all of which they are able to imagine as living prey. But it's not just the movement of the toy they enjoy. What they really seem to want is for me to play the game with them; and they have their methods—a certain cry and way of looking at me—to let me know this is what we should be doing.

So why was I surprised when our pup invented a game? I think because at that time, in the late 1980s—not so very long ago—scientists were still stuck on the question "Do animals have minds?" A cautious search was under way for the answer, and the researchers' caution had spilled over to society at large. In those days, if you suggested that dogs had imaginations or that rats laughed or had some degree of empathy for another's pain, certain other people (and not just scientists) were likely to sneer at you and accuse you of being sentimental and of anthropomorphizing—interpreting an animal's behavior as if the creature were a human dressed up in furs or feathers. My story about Quincie remained that: a story, an anecdote I shared only with close, dog-loving friends. Although I puzzled over Quincie's inventiveness, I didn't know how to interpret her pine cone game or whether to discuss it with the scientists I often interviewed about animals and animal behavior.

Shortly before watching Quincie invent her game, I had another thinking-animal experience—this time with a wild animal, an orphaned chimpanzee, and I was in the company of Jane Goodall, the world's most famous ethologist, a scientist who studies animals as they go about their lives in the natural world.

I had traveled to Goodall's study site in Tanzania, Gombe Stream National Park, to interview her for a biography I was writing about her mentor, Louis Leakey, the renowned fossil hunter who had helped launch her study. While at Gombe I hoped to have some time to watch the chimpanzees, and Goodall thought I should as well. She suggested that I join one of her lead researchers, David Gilagiza, a slender Tanzanian who was then collecting data on mother-and-infant relationships. He would be concentrating on Fifi, a much-revered female in the so-called F-family, and her toddler, Fanni, and infant, Flossi.

Nothing—not the books and articles I had read, or the TV specials I'd watched—had prepared me for my first encounter with wild chimpanzees. Gilagiza and I left the park's guesthouse shortly after dawn and hiked up a narrow trail that led away from the misty shores of Lake Tanganyika and into the woods that sheltered the chimpanzees. It was cool and quiet beneath the forest canopy, and we walked at a steady pace, with Gilagiza stopping now and then to point out plants of interest or places the chimpanzees favored.

The Gombe forest seemed like paradise. Blue butterflies the size of my hand fluttered among the flowers and ferns lining the path, while a tinkling stream sparkled below the trail. I was just about to ask Gilagiza where in these happy woods we would find Fifi when two dark, furry shapes—chimpanzees!—suddenly raced past us. The second one paused just long enough to slap my legs. "That was Frodo, Fifi's son," Gilagiza said, a worried look on his face. "You must watch out for him!"

Frodo, then in his late teens, would eventually become Gombe's dominant male. But when I met him, he was simply an ambitious and frustrated adolescent working his way up the chimpanzee social ladder. Frodo wasn't the smartest or most diplomatic of chimpanzees, but he was strong, and as part of his climb to power he had already beaten up most of the females. Lately he'd begun testing his prowess against human females. He had attacked some of the women researchers—even Goodall—and I should do my best to stay out of his way, Gilagiza said. I nodded, although I wasn't sure how to keep an eye out for this particular chimpanzee. I'd barely caught a glimpse of him and didn't know what I should do if I encountered him again. I also wondered if he would remember me. And if he did, would he try again to impress the other males by hitting me? Were chimpanzees capable of that kind of plotting and planning?

I fell in behind Gilagiza and stayed close—as female chimps often do, joining males that may protect them from other male attackers.

Frodo was my first encounter with a wild chimpanzee, and I wasn't sure what to make of what had happened—or of him. Despite all that I'd read, I had not expected to so quickly meet a thinking chimpanzee. To see that type of behavior, I thought, required weeks and months, even years, of careful watching and note taking. Frodo's slap opened up a host of questions for which I had no answers. Over the next few days, my questions only grew as we spent time watching the chimpanzees, most of whom seemed to ignore us. But that didn't mean that we were like rocks or bushes to them.

Once Gilagiza and I sat near two chimps who were busily stripping the leaves from long, skinny twigs—making tools. When they'd readied these instruments, the chimps took turns dipping them into a small crevice in the earthy mound of a termites' nest and deftly extracting the insects—which they then nibbled as we would peanuts or potato chips. Goodall's studies had shown that this termite fishing requires experience, dexterity, and skill. Why didn't the chimpanzees merely spend their days collecting easy-to-gather fruits? That would be the sensible, machinelike response for any animal hunting food in the wild. I'd rarely thought about how food might taste to wild animals, yet here were two chimpanzees smacking their lips with delight. Could it be that they bothered to fish for termites because they enjoyed this snack? Why wouldn't animals seek out pleasure and fun, just as we do?

On another morning, we watched from a greater distance as two male chimpanzees (neither of them Frodo) brawled through the woods, screaming and slapping at each other. They puffed up their hair to supersize their bodies and uprooted small trees that they shook at each other like spears. I wasn't sure who won or lost this match, but at the end, their wrath spent, one held out his hand to the other—a gesture so easily understood that Gilagiza didn't bother to explain. The two chimps briefly touched, while whimpering, and went their separate ways. Gilagiza didn't know what they had been fighting about, but at the end they patched up their quarrel, just as we do when we know it's better to remain friends with someone than to have an enemy.

We also spent time every day with Fifi's family, usually in picniclike settings under shady fig trees. To my dismay, Frodo, who was Fifi's eldest son, was often with them, but he never looked my way. Had he forgotten our first encounter, or was there no need to put on a show in his family's presence? Whatever the reason, he seemed an entirely different chimpanzee. He feasted quietly with his mother on the sweet fruits or sat with his back to her so she could groom him, while the young Flossi, pink-faced and bright-eyed, swung and tumbled among the vines, as playful as any toddler. We spied on Fifi's family at dusk, too, as they gathered in the treetops and bent the twigs and boughs into leafy beds for the night, then snuggled in together—yet another scene that needed no translating.

Although wonderful to watch, none of these were surprising chimpanzee behaviors. Goodall had thoroughly documented and reported them all. What I didn't expect to witness—just as I hadn't anticipated Frodo's decision to involve me in his social ambitions—was one chimpanzee deceiving another.

Goodall had gone to the "station," a small shed at Gombe, where she had long provided bananas to the chimpanzees to accustom (or habituate) them to people. She handed the fruit to the chimps through a barred window. One afternoon, I watched the proceedings from an adjacent building.

Beethoven, a big male with glossy black fur, was the first visitor. With him was a young female chimpanzee, Dilly, whose mother had disappeared when Dilly was a toddler, leaving her an orphan. As a rule, chimpanzees are raised by their mother, and any orphaned youngsters are cared for by one of their sisters or aunts. But in this case, it was the male, Beethoven, who'd adopted Dilly, Goodall told me later.* Beethoven was both Dilly's benefactor, seeing to it that she had enough to eat from the fruiting trees, and her protector, keeping her safe during any of the group's altercations. But Beethoven's generosity did not extend to sharing bananas.

---

* In 1999, genetic tests showed that Beethoven was Dilly's father, which suggests that the male chimpanzees somehow do recognize their offspring.

Goodall handed Beethoven an armful of the fruits, and he squatted on the grasses in front of the shed and with relish devoured each one. Little Dilly sat close by, watching as each luscious banana slipped down her protector's gullet. Once she reached out a hand to beg, but Beethoven ignored her. Finally, the last banana consumed and his belly full, Beethoven rolled on his back and fell asleep. Dilly sat beside him, grooming his fur.

Goodall had watched the little drama from her window. Unbeknownst to Beethoven, she had held back one banana. When Dilly happened to glance at her, Goodall held up the prized fruit. Normally, a hungry chimpanzee would make a food cry after spotting such a delectable treat. Dilly stifled any sound. She watched as Goodall placed the banana outside the feeding station, away from Beethoven's line of sight. It was as if she and Goodall had exchanged a secret, and like a coconspirator Dilly played her part. She continued grooming the big male, while making cooing, lullaby sounds of contentment.

At last Beethoven began to snore—and Dilly quickly and quietly made her way to the hidden banana. She downed it in three bites. Then she stealthily made her way back to Beethoven's side and resumed her grooming and cooing.

When Goodall and I met up a bit later, I immediately brought up Dilly's behavior.

"What a wonderful demonstration of how chimpanzees can lie and be deceitful!" I said. "Are you going to write that up for a science journal?"

"I can't," Goodall replied.

"But why not?" I asked. Dilly's actions had been so clearly deceptive; she had even connived with Goodall to fool Beethoven. How else could one explain that sequence of events?

Goodall said calmly, "No. Other scientists will say this is only an anecdote and that there is no way to know what Dilly was thinking. If I write it up, everyone will say, 'Oh, Jane, how silly of you. That's anthropomorphizing.'"

She would be attributing a human mental ability to an animal—just as I had done when I told friends about my dog's imagination. But there was a big difference: she was Jane Goodall, an eminent scientist and expert on chimpanzee behavior. If she reported Dilly's deception, I protested, surely other researchers would listen. Goodall and other chimpanzee watchers, such as

Frans de Waal, who had written extensively about the chimpanzees' political machinations, had already established that these apes shared many of the attributes and abilities of humans. The chimpanzees were so similar to us, especially in their expressions and gestures, that at the end of each day, after returning to the guesthouse, I often referred to them as "people" when telling other tourists all that I'd seen.

Goodall nodded. "Yes, it doesn't make much sense to say they aren't thinking or don't have emotions," she said. "Most of us studying animals in the wild see things like this [Dilly's deception] all the time. But we've learned to be careful. We *can* say, 'If Dilly were a human, we would say she was acting deceitfully.'" To say that Dilly—or any animal—had what we would call subjective or personal experiences would be considered unscientific. Although some animals might have an inner, mental life, we had no way of asking them about it and so could not study it.

I asked Goodall how scientists could possibly get around this dilemma. The rules of the game seemed stacked in such a way as to forever preclude knowing what was in the mind of another creature. Goodall agreed but added that because so many researchers were witnessing similar behaviors (and in a variety of species, not just chimpanzees), she thought the science—the study of animal cognition and emotions—would change. "It has to," she said. "It's just a matter of time."

As I listened to Goodall, I realized how little I knew about animal cognition, how scientists define it or study it. Why didn't scientists think it possible to study the thoughts and emotions of animals, particularly one as closely related to us as a chimpanzee? Weren't the chimpanzees (not to mention the baboons that lingered near the guesthouse) thinking? And if they weren't thinking, what were they doing? One look at the baboons, which sometimes hung on the guesthouse's grated windows while surveying my provisions, and I knew what was on their minds: They were waiting for me to make a mistake—to leave a window or the door unlatched—so they could dash inside and steal my food. Simple, common sense, and the baboons' crafty, calculating eyes told me as much. Why did scientists struggle to explain—or simply deny—what seemed so obvious to me?

In fact, we have been wrestling with the question of what goes on inside the minds of animals at least since the time of the Greek philosophers, and surely long before that.*

ARISTOTLE DOUBTED that animals could think rationally, although he did allow that they certainly had appetites and were capable of experiencing sensations, such as hunger, pain, and anger. Stoic philosophers, such as Zeno, had a far narrower view. They discounted the possibility that any animal had thoughts or emotions or sensations and argued that there was no reason to extend any moral considerations to them. Early Christian thinkers, notably St. Augustine, embraced the Stoics' position, and their philosophy has dominated Western ideas about animals ever since.

Most of us are familiar with the Stoics' attitudes because they influenced the seventeenth-century philosopher Réné Descartes. He is most famous for his dualist philosophy, which considered the mind and body as two separate entities—the material body and the immaterial mind or soul. The immaterial part, Descartes argued, linked humans to the mind of God. (It's largely thanks to this Descartian division that many of us in the West don't think of our thoughts or minds as anchored in the physical brain. Instead, we picture them—and cartoonists and illustrators typically draw them this way—as floating outside and above our heads, often in fluffy "thought bubbles," coasting heavenward.)

In Descartes's philosophy, since animals are composed only of material substances, they are necessarily lesser beings. They lack both minds and souls, have no capacity for reason, and are not connected to the mind of God. Instead, Descartes regarded animals as complex automatons, clockworklike things that can see, hear, and touch but are not conscious. Reasoning required language, Descartes argued, and animal calls were only automatic sounds made in response to external stimuli. One of his followers, the philosopher

* This short history is by no means meant as a comprehensive discussion of all the thinkers and scientists who in the past have puzzled over the problem of how or if animals think. For the sake of brevity, I've mentioned only a handful of the many people who have explored these questions.

Nicolas Malebranche, summarized the Descartian view: "[Animals] eat without pleasure, cry without pain, grow without knowing it; they desire nothing, fear nothing, know nothing."

Other philosophers vehemently disagreed with such statements. All one need do, Voltaire practically shouted at Descartes's followers, was to witness one of the vivisectionists dissecting a dog alive (as was done all too often in the seventeenth and eighteenth centuries, to judge from the numerous accounts of this cruel practice). "You discover in him all the same organs of feeling as in yourself," Voltaire exclaimed in 1764. "Answer me, mechanist, has Nature arranged all the springs of feeling in this animal to the end that he might not feel?"

At the time, there was no ready way to settle the matter. On one side were philosophers and theologians who saw humans as the product of a special creation and therefore completely separate from animals. On the other were thinkers who argued that there were sufficient similarities between people and animals that we should be kind and compassionate toward our mute brethren.

In 1859 in *On the Origin of Species*, Charles Darwin presented a solution to this impasse. Animals were not machines or automatons, he argued, but biological organisms that had evolved (and were evolving) in response to natural pressures and changing conditions on Earth. By animals, he meant humans, too. Lumping humans with all other animals was a shocking statement because it meant that we were not specially created. We shared anatomical, physiological, and psychological similarities with animals because, Darwin explained, we were descended from other animals. (Although Darwin did not have genetics at hand to prove his case, we now know that humans and chimpanzees share 98 percent of their genes; humans and fruit flies share 44 percent of their genes.)*

Darwin realized that his discovery of the rules that govern the evolu-

* Despite this evidence, some people, including some scientists, still object to or have difficulty accepting this simple fact of life. One researcher I interviewed, who studies the mental abilities of humans and dogs, told me that "the biggest problem" he faces is that some of his colleagues in the department of human cognition at his university "don't think of humans as part of evolution; they don't fully accept that humans are animals."

tion of life would lead to new understandings in many sciences, including psychology. In *Origin,* he did not discuss in detail how evolution shaped the mental lives of animals, but he foresaw that in the future "psychology will be based on a new foundation, that of the necessary acquirement of each mental power and capacity by gradation." In other words, just as the physical human body had an animal past, so did the human brain and mind. Darwin was convinced this was the case because while writing *Origin* he had collected many observations—some his own, some supplied to him by colleagues—documenting the mental and emotional similarities of humans and animals. He explored both aspects in his next two works, *The Descent of Man* (1871) and *The Expression of the Emotions in Man and Animals* (1872)—books that would help usher in two new fields of science in the twentieth century: ethology, the study of animals in their natural settings, and comparative psychology. (In fact, rereading Darwin for this book, I was struck by how forward thinking he was; his writings could easily serve today as textbooks in cognitive ethology and comparative psychology classes, or as primers for anyone interested in the subjects.)

In *The Descent of Man,* Darwin argued that animals and humans differ in their mental powers only in degree, not in kind—meaning that animals share some of our abilities for reason, memory, and language and also possess an aesthetic sense. Our cognition is more complex than that of other animals, but that complexity is the only difference. All animals, including humans, face the same challenges of life, he observed. They need to find food, mates, and their way around their world, while avoiding predators and hazards—all tasks requiring problem solving and categorizing abilities. Animals use their senses to acquire knowledge about the world, just as humans do, and they act on that knowledge, which gives them a measure of intelligence. Even earthworms, Darwin would argue in a later book, are cognitive beings because, on the basis of his close observations, they have to make judgments about the kinds of leaves they select to plug their burrows. He hadn't expected to find thinking invertebrates and remarked that the hint of intelligence in earthworms "has surprised me more than anything else in regard to worms."

Darwin also placed our human emotions and our manner of expressing them in an evolutionary context. Facial expressions and body postures, he explained, are the signals we and a host of species use to communicate such emotions as joy, fear, affection, and anger. Animals, whether birds, monkeys, apes, or humans, are so strikingly similar in the signals they use to convey an emotion (for instance, when angry, swans puff up their feathers; chimpanzees erect their body hair; we enlarge our chests; we may snarl like a dog, too) that Darwin argued there are "general principles" of expression. Indeed, these emotional signals, including displays of pain, are so useful for social communication that they have become innate instincts, he said—and, thus, are subject to the same forces of natural selection that shape the anatomy of all animals, including humans.

Darwin readily attributed emotions to many species, including lizards, birds, cats, dogs, horses, monkeys, and apes, and wrote of them as beings with rich and complex mental lives. In his writings, animals have motives, intentions, and desires; they do not "seem" to do things—they do things. "Though led by instinct," he wrote about male birds' mating competitions, "they know what they are about, and consciously exert their mental and bodily powers" to try to win the female.

Darwin's protégé George John Romanes, an English evolutionary biologist, took Darwin's study of animal minds a step further. The two were close friends, and Darwin gave the younger man his forty-year collection of notes and papers on animal intelligence. Using these materials and other studies, Romanes compiled a nearly five-hundred-page volume on the subject. Titled simply *Animal Intelligence,* his 1886 book attempted to present all that was then known about the "mind in animals." At the outset, he acknowledged that "we can only *infer* the existence and nature of thoughts" in other organisms on the basis of their behaviors. Nevertheless, Romanes argued that because animals could learn, they must have minds—the same argument that was used at the time to explain the existence of minds in humans.

Following Darwin, Romanes argued that scientists could study the mental states of animals by using a kind of "inverted anthropomorphism"—that is, turning to our human emotions and mental abilities as guides. When we see a dog or monkey (or even a bee) acting in an affectionate or jealous manner, it's likely the animal is experiencing the emotion much as we do (although given its evolutionary distance from us, he noted, much less so in the case of the bee). Romanes then set about examining such topics as the instinctive behaviors of protozoa, memory and sympathy in ants, pugnacity in termites, and jealousy in baboons.

Romanes's book was a hit with the general public but not among other researchers who worried that Romanes and Darwin had both relied too heavily on anecdotes for their evidence and not sufficiently on replicable experiments. Most damningly, Romanes's critics claimed he had done little more in his book than describe animals as fur-clad humans. Although this was not an entirely fair assessment, it continues to be the way that Romanes's work is most often portrayed, ensuring that it is seldom read or that it serves only as a cautionary lesson of how not to do science.

Romanes's writings, even more than those of Darwin, triggered a debate about how scientists should study animal minds. Some thought Darwin's and Romanes's "anecdotal" method had merit if the anecdotes were collected carefully and critically. Others thought a more objective approach was needed; they didn't deny that animals had subjective, mental experiences but claimed that these couldn't be studied because they weren't physically observable and couldn't be measured.

In the early twentieth century, one group of psychologists went even further and refused to accept that conscious activity played a role in any animal's behavior, including that of humans. These scientists called their field "behaviorism," since they restricted their studies to animals' observable behaviors. In their approach, unobservable events—such as desires, memories, ideas, beliefs, emotions, thoughts—were off the study table, both for humans and for animals. "The real question is not whether machines think, but whether men do," B. F. Skinner, probably the most famous of all behaviorists, once said in a sweeping dismissal of mental experience. And if humans do not think but only react to stimuli, why even bother to ask if animals think or feel?

Strict behaviorism, as the science came to be called, had about the same effect on animals as Descartes's philosophy. Once again, animals lost their minds—and any capacity for reason or emotion. Many animal rights observers have noted that it wasn't a coincidence that the use of animals in biomedical studies and pharmacological testing, and as industrially raised and hunted meat and fur products, grew exponentially while behaviorism flourished.

Behaviorism dominated both human and animal psychology in the United States for much of the twentieth century and still has a number of influential adherents in animal cognition research.

Early ethologists were also extremely cautious about the question of animal minds. Niko Tinbergen, the pioneering Dutch ethologist, ruled out attributing mental desires or motivations to what animals do or how they behave. Tinbergen, along with the ethologist Konrad Lorenz and honeybee researcher Karl von Frisch, showed that instinct (meaning innate responses) and imprinting, an instantaneous type of learning, could explain some of the complex behaviors of birds and bees. Their approach had more influence in Europe and England than in America, but the result was the same: for most of the twentieth century, animal minds were off limits for serious scientists, whether studying animals in the wild or in the lab.

Which brings us back to Jane Goodall. She wasn't university educated and had never studied psychology or ethology when she began watching the chimpanzees in 1960. Goodall's mentor, Louis Leakey, thought her lack of training a good thing, since she wouldn't bring preconceived ideas to what she was observing. Unaware that chimpanzees were mindless, she wrote about them in ways that were scientifically forbidden, using words such as *motivation*, *excitement*, and *mood* in her depictions of the apes' behaviors. She described the chimpanzees as having "personalities," gave them "childhoods," and when referring to individual chimps used the personal pronouns *he* and *she* instead of the supposedly objective *it*. (Her editor at the prominent British science journal *Nature* replaced the personal pronouns with *it* on her first manuscript. She replied by crossing out the "its" and restoring the pronouns, later saying that the final version had at least "conferred on the chimpanzees the dignity of their separate sexes.") Her vocabulary cost her dearly at first, because scientists believed such words made the chimpanzees seem like humans and clouded our understanding of their behaviors. Sharply criticized for anthropomorphizing, she was shunned for many years at scientific conferences, and her writings were regarded with suspicion.

But that was years before I met Goodall at Gombe. By then, in 1987, she had won over most of her critics, which made her comments to me about how she could not come right out and say that Dilly had deceived Beethoven all the more surprising. The previous year she had published her great work *The Chimpanzees of Gombe: Patterns of Behavior,* documenting the individual life stories of forty-one of the chimps she had watched for twenty-five years. It was the chimpanzee equivalent of a long-term sociological study of a human community. When you finish her book, you realize why, if you want to understand chimpanzee behavior, you need to know about their gender, childhood, and personalities. Just as with humans, these factors influence how individual chimpanzees make their decisions and relate to their family and community. While Goodall relied heavily on what some might call anecdotes, hers were carefully collected, rich with detail, and substantiated where possible with statistics. She had not spoken to a single chimpanzee, yet she showed us chimpanzees in love, at war, parenting, inventing cultures, and playing politics.

Goodall had said to me that the field would change—and she was right. Her book and studies (as well as those of such prominent ethologists as George Schaller, Frans de Waal, Cynthia Moss, Marc Bekoff, Barbara Smuts, Gordon Burghardt, Louis Herman, and others) helped pave the way for other scientists interested in understanding more about animals than strict behaviorism would allow. By 1987, behaviorism in fact was beginning to wane. As applied to humans, it had started slowly going out of fashion in the 1950s and 1960s when the cognitive revolution, which explains thought and emotion in terms of universal mental mechanisms, swept through human psychology departments. The scientists behind the revolution found it absurd to claim, as the behaviorists did, that human mental experiences cannot be studied because you can't observe them. Physicists also study things that can't be directly observed, noted Bernard Rollin, a philosopher and animal scientist. "[They] talk of all sorts of entities and processes, from gravitation to black holes, which are not directly verifiable or directly tied to experiments," but these "theoretical notions" have not held back their field. In fact, their unobservable ideas "help us to understand the physical world far better than we would without them."

Under behaviorism, everything we humans (and other animals) do was regarded as entirely the result of a stimulus and a response—a dinner bell rang, and you (and the dog) drooled. But as Steven Pinker makes clear in *How the Mind Works,* we know that our beliefs and desires influence our behavior. For example, Pinker says all we have to do is imagine a man who boards a bus because he wants to visit his grandmother and because he knows the bus will take him to her home. The *wanting* and *knowing* are beliefs—mental elements that are weightless, odorless, colorless, tasteless, invisible. Yet the man boards the bus because of them. Although invisible, they cause his concrete, observable actions—but how? The computational theory of mind, which was the key breakthrough of the cognitive revolution, solves the puzzle by picturing the brain as a device for taking in information and processing it. The theory also explains that—and this, as Pinker says, is its brilliance—beliefs and desires *are* information. We don't yet know how this type of information is physically encoded in our brains' neurons, but it is there both causing and affecting our actions and decisions.

So why not apply the discoveries of the cognitive revolution to animal minds, too? Donald Griffin, an animal physiologist who discovered that bats find their way and hunt prey via echolocation, decided in 1976 that it was high time to do so. There was such a "ferment of constructive excitement in ethology," he wrote in *The Question of Animal Awareness,* coupled with the new understandings of how human minds work, that biologists needed to renew the investigation into the "possibility that mental experiences occur in animals and have important effects on their behavior." Griffin had been puzzling about the minds of animals since attending a science symposium in 1948, where von Frisch discussed his experiments with honeybees. "Good God, if mere insects communicate abstract information about distance and direction," Griffin wrote about the moment that "shook up" his thinking about animal cognition, ". . . how could I be so sure that homing birds simply search for familiar landmarks? . . . Although I still considered myself primarily a physiologist and directed my efforts toward mechanistic explanations of animal behavior, I came to see that those mechanisms must be much more subtle and versatile than I had imagined."

Some scientists recoiled from Griffin's book as if Romanes had been reborn. But not so the younger generation, who jumped at Griffin's challenge to find ways via both experiments and field studies to investigate and document animal minds—including the possibility that they might be, as Griffin suggested, self-aware and conscious. Griffin called his new approach "cognitive ethology."

For a field of science to succeed, researchers need a coherent framework—an overarching theory—to guide their questions. Griffin provided that, urging his readers to turn to Darwin and seek an understanding of animal minds through the "evolutionary continuity of mental experience." Without evolution as a guide, the cognitive skills of people do not make sense biologically. What, after all, are the biological roots and history of our cognition? How have evolutionary processes shaped the ability to think and experience emotions? What are the mental experiences of other animals? In short, how and what do animals think?

Griffin, with Darwin in hand, opened a door to these once-forbidden questions, and by the end of the twentieth century, scientists were rushing through.

AFTER WATCHING DILLY'S DECEPTION at Gombe, I'd left with many questions about what we know and don't know about animal minds. I added more questions as the magazines I most often write for, *Science* and *National Geographic*, sent me on assignments to join ethologists and biologists studying a wildlife lover's dream list of animals: elephants, lions and cheetahs, humpback whales, Ethiopian wolves, pink river dolphins, gelada baboons, howler monkeys, golden marmosets, poison dart frogs, and a good half-dozen species of bowerbirds. Each journey was like a crash course in animal behavior—in learning how to watch and think as the scientists do, with an open mind, patience, and an alertness to details.

The elephant watchers I joined in Kenya, for instance, recorded every ear flap of an elephant matriarch and her kin; those subtle movements held the clues to the decisions the elephants were making and wordlessly told the other elephants how they were feeling and what they were about to do.

And in Australia, scientists studying the greater bowerbirds mapped and tabulated the thousands of stones, glass shards, and other decorative bits the male birds use to ornament their bowers—which are like theatrical stages where the males sing and dance to attract females. I felt a surge of pity for the scientist as she knelt next to one bower and showed me the code she'd written on each little stone and piece of glass—there were hundreds, if not thousands, of these, and this was only one bower. Yet the dull, time-consuming work led to the discovery that the bowerbirds aren't just randomly setting out their piles of decorations but arranging them to create the illusion of perspective, a technique often used by artists when painting landscapes. For their illusions, the birds place the largest ornaments farthest away from the opening of their twiggy corridors and the smallest ones closest to it. Thus a female bowerbird standing inside the corridor and looking out would perceive all the items to be about the same size. The researchers proved that the males intentionally create this illusion by rearranging the birds' displays. The birds quickly restored every item to its proper place. Bowerbirds, the scientists concluded, are artists—the first animal, other than humans, that is fully recognized as having an artistic sense.

Over the years, I noticed that many scientists were increasingly at ease talking about the likely mental states and experiences of the animals they were watching—just as Goodall had predicted. Some scientists, such as elephant researcher Joyce Poole, evidenced an almost complete indifference about anthropomorphizing or earlier ideas that denied animals their minds. As we drove through Kenya's Amboseli National Park, Poole addressed the elephants that came up to her Land Rover's window as old friends. When they reached their trunks inside the car to sniff her, she laughed, "Yes, it really is me. And, yes, I know. I've been gone a long time." From her time among the elephants, she knew (and had the data to show) that they had long-term memories and recalled individual elephant friends. They remembered humans, too, distinguishing between those who had never harmed them and those who had heaved spears their way.

In Tanzania's Serengeti National Park, ethologist Sultana Bashir spoke with sorrow about what fate surely lay in store for a male cheetah we were watching. Several months prior to my visit, she and other members of the Serengeti Cheetah Project had placed a radio collar on this sleek cat. We'd driven long hours over the plains, while tracking the collar's *ping*, before Bashir spotted the cheetah among the grasses. He was standing at the base of a rocky kopje—his lookout—and crying in a piteous tone. Sometimes he climbed his rocks to gaze into the distance, other times he paced away, then suddenly swung about and climbed back to the top of his lookout. "Mrrr-roow; mrroowww; mrroww," he called, making a low bleating sound, almost like the cry of a wounded sheep. "That's his distress call," Bashir said. "He's looking for his friend. But I'm afraid he's gone; he was elderly, and I think he's died. Otherwise, he would be here, or nearby."

Male cheetahs maintain large territories that overlap those of several females, and they fight other males—sometimes to the death—to secure their borders. In such battles, it helps to have a friend; indeed, a single male cheetah without a partner is almost assured of losing any fight and all his territory. We sat with the unhappy male until late in the afternoon, and he never ceased his cries. At last, he left the kopje behind and headed off at a trot into the plain's tall grasses. What would become of him if his friend did not return? I asked. "He'll go off to die, I think," Bashir said. "Another male will kill him, or he'll stop eating, get mange—always a sign of stress in cheetahs—and become too weak to defend his territory. Really, he'll die of a broken heart."

I didn't ask Bashir or Poole for evidence to back up their statements. I simply jotted down their words, because the experiences were affecting and because the scientists didn't talk in jargon but spoke openly and simply about what was happening: an elephant had come to visit an old friend; a cheetah was dying from a broken heart.*

In 2006 *National Geographic* asked me to write an article about how animals think. The resulting cover story, *Minds of Their Own*, was published in the March 2008 issue and became the genesis for this book. Reporting took me around the world—from my home in Oregon to several states as well as to Japan, Venezuela, Costa Rica, Australia, Germany, England, Hungary,

---

* Animals and humans are known to develop health issues from stress; they may even die, particularly after losing a mate. There is also increasing acceptance of the idea that social species are badly affected by the loss of a friend or mate. For instance, Laysan albatrosses are monogamous. They nest on Midway Atoll and don't breed until they are eight or nine years old. If they lose their mate, they "go through a year or two of a mourning period," says John Klavitter, a U.S. Fish and Wildlife Service biologist at Midway. "After that, they will do a courtship dance to try to find another mate."

Austria, and Kenya—to meet researchers and their animals. At each lab or field site, I watched raptly as scientists unveiled some aspect of the minds of insects, parrots, crows, blue jays, fish, rats, elephants, dolphins, chimpanzees, wolves, and dogs—and what the animals were thinking.

IT'S PROBABLY BEST at this point to explain what I mean by *thinking*. First, it is an activity that takes place in a physical place, the brain. And second, to borrow from Richard Dawkins and Steven Pinker, the ultimate goal of thinking is to help ensure that the individual with the brain successfully reproduces, thereby making as many copies of the genes that created that brain as possible. What does an animal need to do in order to replicate? It needs to eat, so it must be able to find food. It needs some type of territory or home, so it must be able to find its way through the forest, waters, deserts, or skies, while avoiding hazards. Often an animal will have to elude predators, defend its home turf, or compete with others for a mate. And many animals must raise their young after they are hatched or born.

Animals must learn how to do many of these tasks, and this learning requires them to have memories and the ability to respond to new experiences and new information. The main purpose of learning and memories, and of cognition overall, is to reduce the uncertainties of life and to help an animal predict what may happen in the future.

Thinking in its simplest form may be something like information processing, as scientists such as Alan Turing, one of the key thinkers in the cognitive revolution, suggested in 1950. A brain takes in information via the senses—eyes, tongue, ears, skin, feather, scales, electrically sensitive whiskers, and so on—processes it, and produces a decision in the form of an action or behavior. The action, of course, leads to more information and to another behavior, so that a loop is created between senses, thoughts, and behaviors. Or, as a pair of biologists wrote, thinking "tempers the raw sensory information and prepares new electrical signals to further influence thought and behavior."

Often simple cognition is compared to the set of instructions that directs a computer as it processes data. Of course, that is only a metaphor for how a mind works. In most organisms, the instruction set is much more complicated. Learning, memory, hormones, emotions, gender, age, personality, and social factors—all the messiness of biology—come into play, too.

In human psychology, there's no longer a question about whether cognition operates separately from the emotions. It doesn't. There aren't separate pathways in the brain for thinking and for emotional feelings; they work together on a single track. The same is probably true for all vertebrates, possibly even some invertebrates—even though animals' emotional states are seldom studied. We know less about the emotional side of animal cognition than any other aspect. Comparative psychologists and cognitive ethologists don't deny that animals have emotions, but, as Frans de Waal has pointed out, they have not discovered how to study them.

Many researchers shy from the problem of animal emotions because they worry that such "inner states" cannot be studied—basically, the same argument behaviorists once used as the reason not to study cognition. I've also heard it argued that animal emotions are likely very simple and/or vastly different, even "alien," from those of humans (as if species other than us came from another planet). There is simply no evidence to back up such statements. Because evolution is conservative (for instance, human brains and those of all vertebrates, including fish and amphibians, use the same set of chemicals to transmit signals), it's more likely that many of our emotions *are* similar to those of other animals, as de Waal notes. Why, after all, reinvent sensations, such as fear, pain, or love, and the internal states or mental representations that accompany these? Emotions most likely help animals to survive and reproduce.

When I use the term *thinking*, I'm not implying that animals have language, because thinking does not require language. Thoughts can come as vivid mental images. For instance, the poetry of Samuel Taylor Coleridge often appeared to him in a visual way, and Albert Einstein wrote that his insights typically came as a result of picturing himself doing something like riding a beam of light. Scientists don't know how thoughts are represented in the minds of animals, but some speculate that other animals also think graphically, perhaps in pictures, possibly even animation.

Thinking may or may not require being conscious, depending on how *consciousness* is defined; it's a term that scientists have yet to agree on. In the past, only philosophers studied consciousness. But in recent years, neuroscientists and evolutionary biologists have entered this debate, arguing that because the mind is based in biology, consciousness must be as well—and it must have evolved. "Consciousness does not belong only to humans; it belongs to probably all forms of life that have a nervous system," the distinguished neuroscientist Rodolfo Llinas commented in a 2001 interview for NOVA. He explained, "This is basically what consciousness is about—putting all this relevant stuff there is outside one's head inside, making an image with it, and deciding what to do." Scientists do not yet know how consciousness emerges from the neurons and organization of the brain, but they are making good progress, gaining insights both from neurological patients who have suffered some type of altered consciousness and from monkeys and rats whose brains are scanned while they are making decisions. Several leading cognitive neuroscientists and neuroanatomists are now so confident about the biological basis of consciousness and the idea that other animals are conscious that they wrote a declaration on the subject at a University of Cambridge conference in 2012. It declares, in part, that "humans are not unique in possessing the neurological substrates that generate consciousness. Nonhuman animals, including all mammals and birds, and many other creatures, including octopuses also possess" these—and therefore, they must be conscious. Other mental abilities may be linked to consciousness—such as self-awareness, empathy, insight, and something called "theory of mind," which is the ability to attribute mental beliefs, desires, and intentions to both oneself and others. These, too, must be "evolved, emergent qualities of brains," as the evolutionary biologist Richard Dawkins has described consciousness. And, as such, it is most likely that there are degrees of each one of these abilities in various species throughout the animal kingdom, with the most advanced found in species possessing complex nervous systems and the biology for consciousness.

THERE IS ONE MORE POINT to be made about animal minds and evolution. Evolution is not a progressive force. Although it was once thought that there was a scale of nature or a Great Chain of Being, with all the forms of life ascending in some orderly, preordained fashion—from jellyfish to fish to birds to dogs and cats to us—this is not the case. We are not the culmination of all these "lesser" beings; they are not lesser and we are not the pinnacle of evolution. We are not more highly evolved—either physically or mentally—than our closest genetic ancestor, the chimpanzee.* Nor, despite the belief of many cat owners, are cats more evolved than dogs. Evolution is not linear. It is divergent—which means that we all sit on the limbs of a bushy tree, each species as evolved as the next, the anatomical differences largely a result of ecology and behavior.

The processes of natural selection have shaped every organism on the tree of life in response to the challenges its ancestors faced. Species that haven't succeeded are no longer on the bush; they are extinct. That's why sharks, which have been on earth more than four hundred million years, are considered one of the most successful animals. In comparison, our species, *Homo sapiens,* has been present for only about two hundred thousand years, and it remains to be seen how long we will last. Our human brains are undeniably more complex anatomically than those of sharks. But sharks have survived throughout the ages because they have evolved brains perfectly designed for how they hunt, find mates, and reproduce in their environment.

Although there is no scale of nature and no Great Chain of Being, I've nevertheless organized my book beginning with animals whose brain anatomy is relatively simple and progressing to those that are more complex. I've not attempted to summarize everything that researchers now know about a given animal's cognition. Instead, I've selected specific discoveries that illustrate something new about animal minds and that show how scientists studying animal cognition go about their pursuit and why these researchers are

---

* In 1992, the leading neuroscience journal *Brain, Behavior and Evolution* officially announced the end of the use of the scale of nature in articles discussing the evolution of the brain. It declared that "vague, subjective descriptors such as 'higher' and 'lower' should be avoided" when referring to animal groups.

drawn to their subjects. The book opens with a visit to an ant lab to illustrate how little neural tissue is required for impressive feats of cognitive processing; and it ends with my meetings with wolf and dog researchers, who are trying to tease apart why some of the cognitive abilities of our canine friends are more similar to those of humans than are those of our closest genetic cousins, the chimpanzees. I had also hoped to visit scientists investigating cats' mental talents, but unfortunately very few researchers have looked into the feline mind. Those I spoke with emphasized that cats are bright—they're quick observational learners, for instance—but because cats are independent creatures, getting them to repeat experiments (as is typically required in cognitive studies) is extraordinarily difficult. Immanuel Birmelin, an ethologist at the Society of Animal Behavior Research in Germany, explained how patient he'd had to be in order to run a test to see if cats can count: "One of the cats would do the test once in the morning—only!" he recalled. "Another would do it once in the afternoon—only!" It had taken him four years to show that cats can count to four. Nevertheless, I've added descriptions of studies about cats and how they think wherever possible.

As I wrote the book, I struggled with the use of pronouns, specifically whether to use "who" or "that" to identify an animal. It is standard practice to refer to an animal as "that" but I found myself unable to do this. Alex, the gray parrot, was not a "that"—a thing or an object—any more than was Frodo, the chimpanzee, or Betsy, the language-proficient smart dog. In the end, I settled for a halfway measure, using "who" when writing about known individual animals, and "that" for more general cases. It is not a perfect solution, but it does illustrate the larger question and issues we face as we begin to recognize fully the cognitive and emotional natures of animals.

IF YOU'RE MOST INTERESTED in why our human minds are unique, you'll need to read a different book. I went in search of the minds of animals to better grasp how the other creatures around us perceive and understand the world. What do they think about and how do we know this? Why does it matter? I don't know if knowing more about animal minds will help improve the lives of humans, although this is usually the rationale scientists, particularly neuroscientists, must use to justify their research. But knowing more about the minds and emotions of other animals may help us do a better job of sharing the earth with our fellow creatures and may even open our minds to new ways of perceiving and thinking about our world.

We live at a time when far too many species are either going extinct or are in grave danger of doing so. Many species are dying or losing their homes and habitats, and the resources they need to survive, because of our actions. We are killing many others, from wild fish to elephants, in unsustainable numbers. As these animals disappear, so do their minds. It is a staggering loss, especially when we consider how unique the act of thinking is. We don't yet know of another planet that is as endowed with minds as is ours. There may be other planets with life (such a discovery would not surprise me), but as of now, ours is the only one we know about. Yet only in the last few decades have we seriously attempted to find out what is going on in the minds of our fellow creatures, and we've studied but a mere handful of the many millions of animal minds on our planet.

SO, HOW DO SCIENTISTS PROVE that an animal is thinking? How do they know that they are showing what an animal can do, and not merely doing what humans do so well: projecting our feelings and thoughts onto something or someone else? All scientists engaged in animal cognition studies worry about this, whether they're studying insects, dogs, or dolphins. Our human nature wants to empathize, so much so that we give feelings to our cars and computers. And we ache inside for the poor, lone honeybee we've just used in a cognitive test and must now kill—to protect the integrity of our research—by humanely placing in a freezer. Can we disentangle our thoughts from such emotions and still find a way to look inside another being's mind? Can we really understand the minds of the other animals?

People think about this question—perhaps more often than we let on. Just the other day, my husband and I were out hiking with our collie, Buck. A woman walking a Chihuahua approached from the opposite direction. In spite of the difference in their sizes, our two dogs decided they wanted to meet. They were somewhat wary at first, giving each other sideways glances. Then Buck started slowly wagging his tail, as did the Chihuahua. Watching the two, the Chihuahua's owner asked, "I wonder what they're thinking?"

It's a question many of us have surely asked, and it's the question that drives the scientists in these pages. With new ideas and techniques, they're finally fully exploring what was once one of the most forbidden realms on earth: the animal mind.

1

THE ANT TEACHERS

The brain of an ant is one of the most marvelous atoms of matter in the world, perhaps more so than the brain of man.

CHARLES DARWIN

Nigel Franks leaned over a large square petri dish, studying the ants roaming around inside. Middle-aged and slightly rumpled, Franks has a full beard, graying dark hair, and brown eyes, which on this day were framed by clear safety glasses. Like many scientists, he is single-minded in both work and dress, turning up in his ant lab in button-down shirts and khakis with such regularity that students take note. At least one has written a rap song about him. It begins:

Augh! What?! Augh!
My name is Nigel R. Franks
I study ants
I hang out in my laboratory
Wearing Khaki pants.

Franks had shared the tune with me because it made him laugh, and also because it contained verses that cleverly summarize some of his key observations about ants, the most important being "They're more than the sum of their parts."

"It's not a mystical saying," Franks said about that phrase, as he set up an ant experiment for me to watch in his lab at the University of Bristol in England. "It's the definition of a complex system, which is something that is 'more than the sum of its parts.'" Animal societies, such as ant colonies, are complex systems. An individual ant may seem very simple and inconsequential, but by working together individual ants create highly complex societies. They also solve problems together that they cannot solve individually, although they live in decentralized societies. "They don't have a leader, and they don't have an overview or blueprint of what they're trying to solve or accomplish," Franks said. "So how are they able to form their complex societies? That's what I'm working on with my experiments."

Over the past decade, Franks has produced a steady stream of both groundbreaking and controversial research about how individual ants make the decisions that produce a colony. Decision making in groups, whether ant or human, is increasingly a hot research topic in everything from computer design to election committees to armies, and Franks is regularly sought out as a speaker.

When the ants make decisions, they tend to follow what Franks calls "simple rules of thumb." And the ants' rules, he says, can actually be represented by algorithms—those instruction sets that guide computer programs to predictable results.

Franks and his students decode the ants' decision-making rules and reduce them to algorithms via experiments they conduct here in his lab. They can run multiple experiments every day even in the depths of an English winter, thanks to the room's tall, oversized windows and an overhead array of bright fluorescent lights. Together, the two light sources convince the ants that they're living through an endless summer. Even though it was a gloomy autumn day outside, the ants in Franks's lab were going about their tasks as if it were July—just as those in the petri dish were doing. They had no way of knowing that their peaceful little world was about to change.

"Now you'll see what these people do when their home is destroyed," Franks said, looking up from the ants. For an instant, his glasses caught the light and flashed like those of a wrathful god. He reached into the petri dish and deftly removed the top of the ants' nest—which was simply a "house" of stacked glass and cardboard slides he'd built for them. For the tiny inhabitants, it was as though Jove himself had conjured up a tornado that ripped the roof off their dwelling. Suddenly, their home was open to the elements and all sorts of potential disasters, and the inhabitants swarmed about, rushing to rescue their queen and to save their eggs and larvae. Some furiously stroked each other's antennae, while others clustered together like women and children on a sinking ship, and a few scurried off in what looked to me like pure panic. "No," Franks said, "the people know exactly what to do. They have rules to follow in a crisis like this."

I had only met Franks, who is known as the "Idea Man" among ant biologists for his ingenious experiments, about an hour before the home-wrecking incident. Each time he referred to the ants as "people"—as he often did—I wondered if he'd gone round the bend. I hesitated, then whispered, "They are ants."

Without turning his head, Franks said, "Don't move! Leaning over them like this, we've become their landmarks." I froze in place, but from the corner of my eye I could see him smiling. "Yes, they are ants," he laughed. "I just do that all the time—call them people—because I don't like thinking of them as machines as most people do. Most people think ants are stupid, too, but they're not. They have very sophisticated behaviors, although they don't have language, and they're very generous in giving up their secrets. A marvelous animal."

Franks's ants are cinnamon colored and tiny, about the size of a printed hyphen: -. Their scientific name is *Temnothorax albipennis*, but Franks refers to them more generally as "rock ants" because in the field they live in crevices between the rocks. For Franks's purposes, the rock ants also have the handy habit of willingly changing their homes. They search for and lead their fellow ants to a new nest if their old one is destroyed. Ant species in general will evacuate a demolished nest and send out scouts to look for new quarters, but they don't necessarily use the methods or strategies of the rock ants. Nor is it easy to observe the decision-making process in many ant species, especially those with much larger colonies, whose members can number in the tens of thousands or even hundreds of millions.

In the wild, Franks's ants live in a small area along the rocky coast of southern England. Franks doesn't like to be more specific than that about the ants' location because he worries about the unpredictable nature of people—human people. "You just don't know what someone might do," he says.

Because he's interested in how an animal's environment, lifestyle, and life history affect its actions and decisions, Franks calls himself a behavioral ecologist. But at heart, he is simply a biologist, someone smitten with a love for the study of living creatures. Like many biologists, Franks was first attracted to ants—and animals in general—as a child. He grew up in rural Yorkshire, where his indulgent father let him keep "many varieties of dredged-up pond life, especially insects, and of course, numerous ant colonies." These he stored in his bedroom on the top of his wardrobe cupboard, where he could watch and study them. "Even then, I really wanted to understand how the individuals cooperated to make a functioning whole."

Later, as an undergraduate studying zoology at the University of Leeds, Franks came upon E. O. Wilson's *Insect Societies*. "I read it cover to cover," Franks said. "Afterward, I really never thought about studying anything else. I had no choice. I couldn't imagine being happy doing anything else." When I met him in 2008, Franks estimated that he had been studying ants professionally for more than thirty years.

Some ant researchers are curious about every type and variety of ant and spend their careers learning everything they can about the 12,500 species that are known and trying to describe and catalog the other estimated 11,000 that don't yet have taxonomic labels. But Franks has always focused on a handful of species, preferring to get to know these and their behaviors in depth. For his doctorate, he spent two years in Panama studying the foraging behaviors of army ants—and discovered the challenges of doing experiments in the wild, where clever coatis (raccoonlike carnivores) regularly dismantled—and ate—his carefully designed setups for the ants. "They could reduce you almost to tears some days," Franks said about his battles with the wily coatis. Finally, in the 1980s, Franks settled on the rock ants as the ideal subject to explore what most interested him: how the individual ants form a colony. Franks could do all his experiments in the lab with the rock ants because their colonies are small and because their rock-crevice nests are easily mimicked with the layered stacks of glass and cardboard slides.

Franks collects a few colonies of the ants every year using a "pooter," as aspirators are called in England—basically a device that entomologists employ to suck insects into a glass vial through a rubber tube. "We collect the ants early in the morning before they've started to move, so we get entire colonies," he said.

Franks brings his tiny captives back to his lab, which lies off a long hallway in a Gothic-style building that looks something like the Tower of London. On the door is a small, white, rectangular sign, lettered in black: The Ant Lab. (The name may not be an intentional double entendre, but as soon as I entered the lab, I was struck by the industriousness and—I have to say it—antlike diligence of the scientists inside.) The room itself is cavernous, with high, vaulted ceilings and long black counters with sinks, like a chemistry lab, which it was originally. It's the kind of room where sounds would echo loudly, except that there weren't any sounds to speak of; the Ant Lab was as quiet as a library. At desks and along the counters, Franks's four graduate students (three young women and one young man at the time of my visit) were working on ant experiments or keying ant data into their computers. The students glanced up briefly when Franks introduced me and bent their heads down again to concentrate on their projects. The counter space between the students and their experimental setups was stacked with scores of petri dishes the size and shape of CD jewel cases. Each one was a world unto itself, containing one of the cardboard-and-glass houses and a colony of about two hundred individual ants.

"I have everyone in their society," Franks said, referring to the population of each petri dish. He paused for a beat to allow this fact to sink in. "If you're studying apes or monkeys or birds in a lab," he added, "you only have a portion of their society. You never have *everyone*—every single individual."

It is usually only biologists studying animals in the wild who have intact societies to observe, although researchers have made many discoveries watching captive colonies of birds, fish, chimpanzees, and other species. Whether in the wild or in a lab, knowing each individual is the key to unlocking and understanding how and why animals behave as they do. It's not always easy for us to tell animals apart, but scientists have come up with some surprising markers: every African lion, for instance, has its own pattern of whisker spots; the underside of each humpback whale's tail fluke bears a unique black-and-white Rorschach-like decoration; and every elephant's ears are frayed and cut distinctively along the edges.

Ants lack such obvious markings, although scientists say that in some species ants do know each other as individuals. "They use chemical odors—pheromones—to recognize and communicate with each other," Franks said. Using the scent receptors on their antennae, they can tell if another ant is friend or foe, spread warnings of attacks, call for recruits, and pass along news of the hunt. They also map their world with scent by lightly touching their gasters—an ant's hindquarters—to the ground and extruding a pheromone from a gland, leaving trails for their sisters to follow. (A quick note to readers who didn't have the joy of owning an ant farm as a child: in ant colonies, as a rule, the worker ants are the females and all are daughters of the queen. It's her job to reproduce. Instead of breeding, the workers feed their queen-mother and help rear her offspring. The queen also produces winged-male drones, but the drones never work. They are cared for by the workers until the beginning of a new mating season, when they fly away. They will mate with a winged queen, who stores the sperm from one or more males in her reproductive tract and then finds a suitable place to establish a new colony. There she breaks off her wings and begins laying eggs that soon develop into female workers. The males die shortly after mating.) In most species of ants, an individual doesn't recognize a trail she's made as being her own, but, for reasons not yet clear, each rock ant can distinguish her own pheromone trail from those of her nestmates.

Of course, the ants' scents are beyond our ken, and when we look at a colony, we simply see ants that all look alike. And ants the size of Franks's appear at first glance to be nothing more than busy, brown punctuation marks. So Franks bestows individuality on them by painting colorful patterns of dots on their heads, thoraxes, and gasters. "It's an enormous labor to do this," he said. "It takes ten hours to paint the full colony." He paused, thinking of the stacks of unpainted colonies that sat on the lab's counters. "It's a horrible job, really." Franks shook his head, when I asked how many ants he had decorated over the years. "Too many," he said. While his students now shoulder most of this task, Franks also occasionally paints some of the colonies.

Later that morning, he showed me how to paint an ant. With a pair of stainless steel forceps that are so thin and flexible they do not hurt the ants, Franks carefully picked one up, seizing her just below her thorax. He deftly slipped her headfirst into a narrow slit in a piece of rubber foam placed beneath his microscope. To keep the ant from squirming, he wafted carbon dioxide over her, which temporarily sedated her, then dropped a tiny dot of paint on her gaster from the tip of a wire that was as thin as a human eyelash. "You don't want to get the paint on their eyes or antennae," he said, "and you don't want the paint splashed all over their bodies. You want just the most minute, little bit of paint. So, you just sort of let it go . . . *splat*," he said, splatting another drop of paint on the ant.

Each ant in a colony is daubed with a unique pattern of four hues of paint. So an ant may be given a royal blue dot on her head, one dot of white and another of true red on her gaster, and one of lemon yellow on her thorax, while another ant is marked with green on her head, blue and yellow spots on her gaster, and a red dot on her thorax. When they're all painted and back at work in their petri dishes, the ants look like tiny, enameled jewels, scurrying purposefully about.

Before he wrecked the ants' home, Franks handed me a magnifying glass, and through it I studied the painted ants as they busied themselves in their walled landscape. Franks had scattered bits of sand and set out fruit fly carcasses (which, compared to the ants, looked as large as mammoths) and tiny, doll-sized aluminum dishes filled with water and honey as provisions. Magnified, the painted ants looked less like jewels and more like Elizabethan actors or courtiers fashionably dressed in pantaloons and fancy hats. They moved around their minilandscape, where even the grains of sand dwarfed them, with sure steps, harvesting supplies and working together to carry their bounty back to the nest.

"If they didn't have the little paint spots, we wouldn't see them as individuals," said Franks. "The colors and patterns change our whole outlook on them—which is very intriguing. We humans respond so differently when we see animals as individuals." No one in his lab has named any of the ants yet (although some of his students think that the colonies have personalities, since the ants in some nests are exceptionally fast at their work, while the ants in other colonies seem to be slower or more hesitant). But—and this is key to

Franks's success in decoding his ants' behaviors—it is only after the ants are painted that they become "people" to him. "You stop looking at them as a colony, as a kind of single, superorganism," he explained, "and start thinking: 'Well, what are these individuals doing?' We know each one is making her own decisions, that there isn't centralized leadership."

An unmarked colony remains a "black box," Franks added, one that we cannot see into and cannot understand: even though it is clear that the colony's members are communicating, and evaluating each other's contributions to the group effort, we have no idea of how they achieve their goals. But a colony of painted ants is no longer opaque, because each ant can be identified as an individual, and her actions and decisions tracked and recorded.

Since Franks is not an ethologist, he doesn't track the ants to see what they are normally doing. He wants to understand how they solve a problem, such as deciding what to do after a simulated natural disaster, like the roof-wrecking incident. He calls these events "challenges."

How do the ants respond, for instance, if the scientists waft the scent of an enemy colony over their nest, or sow havoc with a windstorm created by an electric fan? Do the ants immediately abandon their nest to move to a new house? What if their old nest is better than any other home they can find; in a disaster, do they still move? And since they do not have a leader, how do the ants agree on one course of action?

Depending on the experiment, Franks provides the ants with one or more new nests to move to. He then watches and records the ants' behaviors as they abandon their old nest, search for and choose a new one, and rebuild the colony. "By destroying their old nest in these experiments, we're basically taking apart the colony and then watching them reassemble it," Franks said. By doing this numerous times with many colonies, Franks can say exactly what decisions the rock ants will make when confronted with the challenges they often face in nature.

I HAD COME TO FRANKS'S LAB because in the course of asking questions like these he had discovered that his rock ants teach. Moreover, he and his coauthor Tom Richardson (who worked on this topic as part of his master's thesis with Franks and is now at the Swiss Federal Institute of Technology in Zurich) argued in *Nature* that their ants were the "first non-human animals" to qualify as teachers. To claim that an ant about the size of the head of a pin can teach is almost as daring as Copernicus's claim that Earth was not the center of the universe. Scientists have a long history of debating whether any animal teaches. Some argue, for example, that mother cats teach their kittens how to hunt and that rat dams instruct their pups to avoid toxic food. Both lessons seem very sensible things for the mothers to teach. And if you read the papers announcing these discoveries, you would likely find yourself nodding in agreement: yes, definitely, cat and rat mothers really have their lesson plans squared away. But other researchers often seem to regard such findings as challenges and, like scoffing parents (you think your kid is *so* smart), delight in deflating their colleagues' balloons. Only a few months before Franks published his and Richardson's study on the tutoring ants, another group led by Bennett Galef, a comparative psychologist at McMaster University in Canada, announced in *Animal Behaviour* that, in fact, rat mothers "do not teach their young what to eat," despite how beneficial (and smart) that would be—particularly in a world where rat pups are sure to find plenty of tempting piles of D-Kon kibbles lying about.

Still, although it was easier for me to imagine a researcher designing a teaching experiment for a rat (a vertebrate with a backbone and internal skeleton, like us) than for an insect (a six-legged invertebrate with an external skeleton), Franks's idea that ants teach each other fit with a wealth of studies over the last decade showing that insects' cognitive abilities are surprisingly rich. Scientists have revealed, for instance, that social wasps recognize each other's faces, meaning that the wasps know each other as individuals, and that female field crickets, like experienced bachelorettes, remember the courtship songs of the males and use this social information when choosing a mate; honeybees can learn to categorize things according to whether they are the same or different, an indication they have some understanding of abstract concepts; honeybees can also learn to discriminate among human faces (a talent once argued as belonging solely to humans and necessary for developing our social bonds); moths remember what they ate when they were caterpillars; some ant species have a sense of their mortality and change their work roles as they age; and fruit flies exhibit enough spontaneous behaviors when tethered to a pole and deprived of any sensory input that the researchers behind this experiment concluded the flies have some semblance of free will. Crickets, water striders (those long-legged insects that seem to walk on water), tent caterpillars, and fruit flies (and no doubt many more as yet untested species of insects) have all been found to have personalities—meaning that some are bolder in their actions than others.

The minuscule brains of insects are composed of ganglia, collections of nerve cells. Within the brain's ganglia are two other neuronal structures, called mushroom bodies because of their cap-and-stem shape. The mushroom bodies seem to be important for learning and forming memories. Although some insect brains (such as those of the rock ants) may weigh less than a single grass seed, they are now thought by many researchers to have striking similarities to the mammalian brain's cortex in design and function. Further, the size of such minuscule brains is far less important than most of us think. What is true for computers also holds for animals: a big machine isn't necessarily a better one, nor is a big brain a good measure of one's ability to think or solve problems. More important, it seems, are the number of neuronal connections, and insects, particularly the social ones such as ants and bees, have very dense neural networks. Indeed, the brains of social wasps and honeybees—and very likely ants—increase in size as they take on more important tasks in their nests.

This is not to say that insects are hard-shelled, miniature humans; they have innate behaviors that limit their actions and responses. But these are more flexible than researchers would have suspected even a decade ago. For much of the twentieth century, when behaviorism ruled the study of animal cognition, scientists viewed insects (and almost all animals) as hardwired beings—their brains as fixed as the wiring on a circuit board. Each neuron in an insect's brain (and neurons in ants can number over one hundred thousand; we have about one hundred billion) was assumed to be welded to another neuron (thus hardwired), and all were set in an unchangeable pattern from the moment an insect hatched until it died. Insects' behaviors were considered to be as immutable as their brains; they functioned solely by built-in "instinct," a term that many researchers have criticized as not being very well defined but that implied that insects were essentially slaves to their genetic instructions and unable to change their reactions. They supposedly learned nothing from any experience, no matter how useful, or good or bad.

Even for those who thought otherwise, it seemed difficult if not impossible to find some definitive experimental way to get inside an insect's mind. One zoologist and psychologist, Vincent Dethier, spent eighteen years, from 1948 to 1966, at the University of Pennsylvania attempting to prove what he felt must be true: that blowflies (*Phormia regina*) can learn; that is, they can alter their behavior because of something they experienced.* "Perhaps" the flies "are little machines in a deep sleep," as most people think, he wrote in 1964, "but looking at their . . . staring eyes . . . one cannot help at times wondering if there is anyone inside." Ultimately, after trying everything he could imagine to demonstrate that flies could learn, Dethier gave up. "You name it and we've tried it," he told the *Washington Post* in 1966. The paper reported on this failure by printing a horror movie–like photo of a fly's magnified face with the caption "Can't learn anything."

Five short years later in 1971, one of Dethier's former students, Margaret Nelson, finally showed that blowflies *can* learn by teaching them that if they dunked their legs in water, they would get a tasty drink of sugar water. Her study, along with Karl von Frisch's decoding of the honeybees' dance, marked a turning point in our understanding of insect cognition. Not long after Nelson's study was published, others showed that fruit flies also learn and have memories; scientists have now isolated the fruit flies' genes that underlie these abilities. Indeed, all animals and even bacteria can learn from

---

* Charles Abramson, a comparative psychologist and historian of his field at the University of Oklahoma, believes that a black American scientist, Charles Henry Turner, devised many of the first experiments showing that a wide range of insects—ants, bees, cockroaches, and spiders—can learn. Abramson regards Turner's research as "the foundation of all subsequent insect-learning studies." Many of Turner's papers are cited by scientists today, although they were published in the late nineteenth and early twentieth centuries. Turner earned a PhD in zoology from the University of Chicago but was never hired at any university, despite applying for numerous professorships. He taught high school science courses and carried out his groundbreaking studies—three of which were published in *Science*—in his spare time.

experience, just as humans can.* There seems to be an "innate schoolmarm" in all of us, as Konrad Lorenz noted about this predisposition for learning. It's a handy ability to have because, although the world is often unpredictable, animals are still likely to encounter certain situations repeatedly. Remembering how to react—"Here's some water; if I put my feet in it, I'll get some sugar!"—can pay handsomely.

BUT IF ALL ANIMALS ARE STUDENTS, capable of learning something about their world even in the briefest of life spans, not one—not even a chimpanzee—has been fully celebrated as a teacher. So when Franks and Richardson published their discovery about teaching in ants there was a storm of protest. Human psychologists and many (although not all) of Franks's colleagues studying insect cognition bristled at the idea that ants could do such a thing. His former mentors, famed ant researchers E. O. Wilson and Bert Hölldobler, dismissed Franks's image of tutorial ants as nothing more than a "charming metaphor."

When Darwin discovered that earthworms had "some degree of intelligence" because they made judgments about the best material to use for blocking their tunnels, he at least added, "This will strike everyone as very improbable." But Franks and Richardson did not hedge their remarks, and in the years since the publication of their paper they had not wavered about their interpretation of the ants' behavior. What caused Franks to stand by their assertion, I wondered. What made him so convinced that his ants were teachers?

* In 2008, scientists at Princeton University discovered that bacteria can learn and anticipate future conditions, just as Pavlov's dogs learned to anticipate food when a bell rang. In the researchers' experiment, colonies of *E. coli* developed the ability to associate higher temperatures with a lack of oxygen; in response, they lowered their metabolism. "Associative learning in dogs and humans happens over the course of the organism's lifetime and involves modifications to the strength of connections between neurons in the brain," says Saeed Tavazoie, the lead scientist of this study. "The learning that we have discovered occurs over a long evolutionary time-scale and involves changes in the connections between networks of genes." Nevertheless, the same fundamental principles of associative learning are at play.

Franks thought the best way to answer that question was to let me watch ants in the act of teaching, but first he thought I should understand how cognitive scientists define teaching. "There was a definition of teaching in nonhuman animals that everyone accepted" when he and Richardson began their experiment, he said. "It set out certain criteria that must be met before an animal could be said to be teaching. So we looked at the ants' behavior to see if it fulfilled those criteria. Does an ant 'modify its behavior in the presence of another, at a cost to itself, so that another individual can learn more quickly?' And the answer is, yes, ants do this."*

Franks sat with his eyes half closed and his eyelids fluttering as he recited the definition. He clearly had high expectations for his ants, I said.

"My approach has always been to investigate the procedures the ants use to solve problems," he replied. "They may not have thoughts. But they gather information, and they act on it. And a good part of their decision making requires social communication."

As he daily watched the ants in his lab during various experiments, Franks took careful note of those social communication moments. He grew particularly intrigued by how an ant who had found some tasty treat or a better home site conveyed that news to one of her sisters and then showed her the path. More was involved than merely following a chemical trail, Franks decided, but what?

TO ANSWER THIS QUESTION, Franks launched a series of experiments similar to the one he was showing me. The simple act of destroying a colony's home triggers the instant response I'd witnessed when Franks removed the

* Tim Caro and Marc Hauser, two well-known animal cognition researchers, developed the definition for teaching "in nonhuman animals" in 1992 that Franks used for his experiment: "An individual actor A [the *tutor*] can be said to teach if it modifies its behaviour only in the presence of a naïve observer, B [the *pupil*], at some cost or at least without obtaining an immediate benefit for itself. A's behaviour thereby encourages or punishes B's behaviour, or provides B with experience, or sets an example for B. As a result, B acquires knowledge, or learns a skill earlier in life or more rapidly or efficiently than it might otherwise do so, or would not learn at all."

roof of their nest: Find a new home! "They must get their larvae and queen out of the glare, and back in a safe, dark crevice," said Franks. Franks had arranged another such home—also made of glass and cardboard slides, and with an ant-sized doorway cut in the cardboard—for them in a second petri dish about eighteen inches away from their old one. Both dishes were placed inside a larger, plastic arena. Franks had connected the two petri dishes with a strip of 3-inch-wide acetate tape, fixing it so that it arched like a bridge between the dishes. Franks jokingly called the tape the "Millennium Bridge," as in the new bridge that spans the Thames in London. The ants' Millennium Bridge presented the only path out of their old dish with its broken home and into the new one. It was up to the older, foraging workers—the scouts—to find the bridge, which they did in a matter of minutes. "They do have knowledge about their world [the furnishings inside the petri dish] from their previous foraging," said Franks. They didn't waste time looking for a new home in this familiar territory but seized the opportunity the bridge afforded to head into the unknown. Nor did they lay a single trail or form a column of ants, as many other species do. Instead, they moved independently, like true explorers. "Their job is to find a new home," Franks explained, "and as soon as they find the house in the other dish, they'll investigate it to see if it meets their criteria."

So these little ants were real-estate appraisers, too. The image made me smile, but Franks and his colleagues' research on how rock ants select their nests is widely praised, even by the teaching critics. "Yes," Franks said. "In fact, they are very discriminating consumers, and make sophisticated decisions when choosing a new home."

Through their experiments, Franks and his students have compiled a lengthy list of what these ants expect in a home, including the necessary amount of floor space and ceiling height, and the number and width of entrances. The ants prefer a nest that has an internal area of about 2,000 square millimeters, is 2 millimeters deep, and has a 2.5 millimeter-wide entrance. "It's extraordinary how precisely they measure things," Franks said. A nest should be dark, too, and clean. Scouts will spurn even the finest manor if there are ant corpses lying about or if they have nasty neighbors.

Now several more scout ants were crossing the bridge, each one laying her distinctive chemical trail and, Franks thinks, learning nearby landmarks. "Their eyes are on the top of their heads, and they see above and to the side," he said. "They seem to use any vertical marker, such as you and I are now. It's similar to us using a church spire or other tall landmark to find our way." In the wild, they must register the differences among the grasses and shrubby bushes that shoot skyward over their heads, as a human would when seeking a path in a redwood forest.

About five minutes after their home was wrecked, one scout had already found the new home, and less than one minute later, the mouse hole–like entrance, too. She didn't hesitate but crossed the threshold, vanishing from our view. Franks nodded his approval.

"She's checking now to see if it's a good home," he said. Another scout also found the new nest and entrance and made her way inside. Shortly afterward, the first one reappeared and without stopping hurried back across the Millennium Bridge to her old home. When she met another worker ant, the two briefly touched their antennae. The scout then started back to the new nest, with the worker following close behind. "The scout will teach her sister the route to the new nest," Franks said.

The pair of ants began what ant researchers term a "tandem run," because like riding a tandem bicycle, it's a task for two.* They moved slowly over the bridge with the leader taking a few slow steps, then stopping to wait for her follower to catch up. It was a slow, start-stop journey, but eventually the pair reached the new nest, and the follower went inside—but only for a brief survey. Her "teacher" had already made a beeline back to their old home, and this second ant returned there quickly as well. Soon, they were both engaged in new tandem runs, with each one guiding another ant to the new home. "They'll continue like this, with one ant enabling another to find the new nest, until they reach a certain number of ants who know the way."

At that point, the workers who have learned the route will switch to carrying the remainder of the ants, who are usually the younger ones, as well as the queen, larvae, and eggs. "Carrying is much faster—actually, three times

* Although the tandem runs were first described in 1896 by Gottfrid Agaton Adlerz, E. O. Wilson gave the behavior its name.

as fast as doing a tandem run," said Franks, but they don't use the carrying method until they've attained a specific number of ants who know where the new nest is.

To carry another ant, a worker grabs her by the mandibles and hoists her over her back, then runs off, seemingly indifferent to her burden. An ant being carried looks downward; she cannot see the overhead landmarks and so does not learn the path between the old and new houses. Instead, she helps inside the new home, arranging the larvae and eggs into the same positions they had in their old nest. "It's an exact replica of their old home," said Franks. "Everything and everyone has a specific place to be."

With the ants' emigration well under way (it would take three hours for them to complete their move), Franks suggested we watch close-up videos of a few tandem runs on his office computer. He videotapes every experiment and later replays these numerous times to study and record each ant's behavior.

When the ants were magnified on the computer's screen, it was easier to see what Franks called teaching. He chose a few clips of specific tandem runs and pointed out what the two ants were doing during their start-stop journeys. "See, the leader is walking very slowly. She won't go forward until her pupil taps her on the leg and gaster with her antennae." The leader, in this video, had a white dot on her gaster, while her student's was marked with lipstick red. In my notes, I called them "TW" and "PR," for Teacher White and Pupil Red. Franks explained that Pupil Red didn't simply follow her teacher's chemical trail to the new nest but had to learn the landmarks along the way. That was why she moved so slowly, stopping now and then to swing her head like a cow, left and right, almost as if searching for clues. "She's learning the landmarks," Franks said, adding, "We don't yet know how much they depend on landmarks versus path integration—which is something like the dead reckoning sailors do—but they probably use both."*

After Pupil Red had learned a section of the trail, she tapped her antennae against the legs of Teacher White, who responded by moving forward

* Sailors use dead reckoning to find their way by recording the distance and direction of each sequence of their journey. That way, they know where they are relative to where they came from.

a few steps. But whenever her student stopped tapping, Teacher White also stopped walking and waited.

"Now, here's the surprising part," Franks said. "Watch what the leader does when she doesn't get that tap." Teacher White had been waiting for almost thirty seconds, while her pupil lingered several steps behind. "Her student is lost," Franks said. "She's missed part of the trail." As I watched her on the screen, I saw that Franks was right. The lost ant rubbed her antennae with her front legs, swung her head, took an uncertain step, and looked decidedly . . . well, lost.

And like any good tutor, Teacher White waited patiently for Pupil Red to catch up and learn this part of the path. The pair then resumed their journey. This was why the tandem run between the two nests was so painfully slow. In fact, it wasn't a "run" at all. It was more like a memorization lesson while walking (like simultaneously walking and memorizing the skyscrapers of Lower Manhattan), with a teacher right there to monitor her student's progress every step of the way.

For Franks, the leader ants in these tandem runs behave exactly like the hypothetical teacher in the definition. The leader *did* modify her behavior—only when she had discovered a nest or food that her nestmates needed to know about did she produce a special pheromone that told them to follow her. And she moved in this slow manner only when she had another ant in tow, following her. The teacher also endured a cost, since she could have moved between the two sites far faster without a pupil following her, leaving her more vulnerable to predation or some other mishap. The teacher was also sensitive to her student and would wait a minute or longer for the student's tap, signaling "I've learned that."

"They do teach, and they teach with evaluation," Franks said, gazing at me steadily, while lifting his chin ever so slightly and firmly closing his mouth—one of those "special expressions of man" that Darwin took note of when documenting the continuity of emotions in animals and humans. Franks was proud of his ants.

Earlier, as we watched the ants steadily go about the task of relocating their colony, he had said that he "really admired them, because they are so tenacious; they are just so determined to get on with the job. It's like seeing a person who has many difficulties to overcome and who just carries on. There's something very admirable about that."

Franks's admiration and affection for his ants make it impossible for him to throw out any of the old colonies ("I do confess to a bit of sentimentality for some of these colonies"), although many of the experiments require that he use new ones. So, he and his wife, fellow ant researcher Ana Sendova-Franks, safely tuck the ants in their petri dishes into shoeboxes and take them home. They store the colonies in their garage and care for them, replenishing their supplies of food and water. Their garage now holds so many shoebox ant-condos that Franks said, blushing, "We certainly can't get the car in anymore. But I like it that the ants come home with us."

And even while stored in the Frankses' garage, the ants teach them new things. By continuing to care for them, Franks and his wife have discovered that the workers can live to age six, and the queens much longer. And from that discovery, Franks and his students have devised new experiments to investigate such questions as: How does an ant know what to do at the various stages of her life? Can a young ant, a *callow,* as a newly pupated ant is called, lead a tandem run? Does she need to have participated in one first, or can she rise to the task if all the older ants are gone? Which behaviors are instinctive, that is, influenced by genes, and which ones are learned?

"We have many questions to ask," Franks said, "and so we carry on."

THE DEBATE ABOUT FRANKS'S DISCOVERY of teaching in ants is far from settled. Some researchers argue that the ants don't teach because they don't improve the skills of their students; others sniff that the ants are merely transferring information, much as we do when we tell someone about the location of a good restaurant. And then there are those who have decided that the best response is simply to further refine the definition of teaching, adding to the earlier criteria a new one. According to this new, modified definition, a teacher needs to know the student's state of knowledge or ignorance—something exceedingly difficult to test in a species without language. Such a stringent definition would automatically exclude ants and all other animals, other than humans. "There's something rather sad and disappointing about that [reaction]," Franks said, especially since such new, exclusionary definitions often bring the research to a halt. The "pleas for changes in the definition of teaching," he and Richardson wrote in a subsequent paper, "seem to be tracking our own understanding of what is special when humans teach." Our understanding of teaching, they argued, would be better served "by focusing on the underlying similarities among different" species that may be doing something like teaching. Such an approach would enable researchers to "avoid succumbing to the understandable temptation to use the most exotic, extreme case [of teaching], i.e., the human one, to define what is perhaps a relatively common phenomenon."

But not all scientists have reacted defensively to Franks's study. Many have found it to be inspiring. "It's reinvigorated the field and led to a mini-explosion of studies about animal teaching," Kevin Laland, a cognitive biologist at St. Andrews University in Scotland, told me in a telephone interview. He realized after reading Franks's paper that the stickleback fish he studies do something very similar to Franks's ants. Laland and others argue that the old definition, written in 1992, that Franks used to test his ants' teaching skills remains the best for assessing the tutoring abilities of other species. More is to be learned, he says, by making a list of all the species that scientists say teach or do something akin to teaching as described by that definition. When he and his colleagues did this, "we ended up with a very bizarre distribution: ants, honeybees, cats, meerkats [squirrel-sized, social carnivores of the Kalahari Desert in southern Africa], pied babblers [pigeon-sized, social birds, also of the Kalahari], even chickens."

It's the kind of list Darwin would have applauded, because it suggests that teaching evolved independently several times, and because it challenges the old idea that higher cognitive skills must evolve rung by rung, up a ladder, from lower to higher species. It also opens the door to other questions, making teaching a subject for further scientific study. What do these animals have in common that would lead them to do something like teaching? Are there certain behaviors or ways of living that make teaching more likely to evolve?

Laland thinks so, pointing out that all of these species are highly social and live in closely related, extended families, where social learning occurs, sometimes inadvertently and at other times through imitation and coaching.

For instance, to survive in the dry Kalahari Desert, meerkat pups must learn how to catch, kill, and eat poisonous scorpions. Without guidance from an adult, a pup's first scorpion meal could also be its last, since a meerkat pup can die from a scorpion's sting. To prevent a pup from being harmed, an adult meerkat bites off the scorpion's tail and wipes the tail-less body on the sand to remove any trace of venom. The adult then gives the live, disabled meal to the pup, which can then play with its dinner without being injured. As the pups grow older, the adults toss them scorpions that aren't as disabled. Adults also nudge prey at the pups, retrieve scorpions or other prey that try to escape, and further injure any scorpion that's giving junior a bit of a struggle. "It's probably the best example of teaching that's been documented so far," says Laland, who expects that teaching in various forms may be very common. "We shouldn't expect teaching in other animals to look exactly like the teaching we humans do," he cautions. "But we shouldn't be surprised to find some forms of it—whether in the form of imitation, or coaching, or transferring information—throughout the animal kingdom."

Even in ants the size of hyphens.

ON MY LAST MORNING in Franks's lab, he offered one final thought about his ants—and about thinking in general.

When I stopped by the lab, Franks was watching a new move-your-house experiment unfold with one of his graduate students, Elizabeth Franklin. She had used a gentle brush to sweep away all the elders in this colony, leaving behind only naive ants, who had never been part of a tandem run. Some seemed to know what to do anyway and were guiding ants to the new estate; others were in too much of a hurry and had left their slow pupils behind; while a few simply decided that carrying their sisters was the best strategy. All were busy.

Franks sat hunkered at the edge of the arena, his chin in his hands, his eyes fixed on the ants. He was chuckling. "Sometimes they just look so purposeful. Look at her," he said, pointing to one that was racing back to the original, broken home. "She's led one ant to the new nest, and she's off to get another, and she's just hurrying along. It's almost as if they get a buzz out of doing something successfully."

Franks was silent for a moment, watching the ants he loves. Then he looked up again. "I would never say that these ants are thinking," he said, arching his eyebrows for emphasis. "But that's what intrigues me—because in many ways, they behave *as if* they are thinking," that is, as if they were consciously reflecting on their actions and knew what they were attempting to accomplish. "They've taught me that very sophisticated behaviors don't necessarily need to involve thought or language or theory of mind," the ability to understand what another person is thinking or feeling.

"The ants provide a cautionary note to us" about our own thought processes, Franks continued. "Could it be that, like the ants, all we have are algorithms in our heads, and we just put a gloss of reasoning on these? That all we're doing is saying to ourselves, 'I think I thought it through'?"

It was a humbling thought. Are our brains really working at a higher cognitive level, or do we just think so because we have a bigger store of instincts, experiences, and memories and have a constant mental monologue, reviewing and revising what we just did?

"Ant colonies are no different from brains in many respects," Douglas Hofstadter, a computer scientist, wrote some thirty years ago in *Godel, Escher, Bach: An Eternal Golden Braid.* He pointed out that a form of higher-level intelligence emerges in both ant colonies and brains from groups of "dumb" beings—ants and neurons. Are ants and neurons really so similar? After quoting Hofstadter's words in his new book, *Honeybee Democracy,* the insect sociobiologist Thomas Seeley suggests that "it is possible that primate brains and honeybee swarms [and ant colonies] have independently evolved the same basic decision-making scheme precisely because it provides a good approximation of optimal decision-making." If so, he continues, then this is an "astonishing convergence . . . of two physically distinct forms of a 'thinking machine'—a brain built of neurons and a swarm built of bees."

But the brain built of neurons, which is found in all vertebrates, does seem to do something more than the bees' swarm or the ants' colony. In its decision making, a vertebrate's brain has to contend with and incorporate those elusive sensations we call emotions. Why is this? How can they possibly help an animal to think, and to reach those optimal decisions?

It was time to visit some vertebrates, and the scientists investigating their minds.

2

AMONG FISH

I wouldn't deliberately eat a grouper any more than I'd eat a cocker spaniel. They're so good-natured, so curious. . . . Fish are sensitive, they have personalities; they hurt when they're wounded.

SYLVIA EARLE

What is it like to be a fish? Or to be a bird, bat, bee, or tiger? The scientific consensus holds that we will never know exactly what it is like to be any animal; as much as we would like to, and despite all of our intensive studies, we can never fully understand another animal's experience of the world. And yet, when the scientists at a 1991 international gathering of ethologists were each asked why they had elected to study a particular species, they overwhelmingly responded that their primary motivation was the desire to know what it was like to be that animal.

For Stefan Schuster, a neuroscientist at Germany's University of Bayreuth, the urge to enter another species' mind—specifically, the mind of fish—grabbed him as a child, and it has never loosened its hold, even now that he's in his midforties and has spent nearly twenty years investigating how fish think and make decisions. His journey to understand the minds of fish has taken him into the very cells of a fish's brain where the decisions of life and death are made. Along the way, he's made key discoveries about the sophisticated mental abilities of archerfish, the sharpshooters of the piscine world.

In the wild, archerfish inhabit the mangrove swamps of Southeast Asia, Australia, and many Pacific islands. They hunt by firing precisely aimed blasts of water from their mouths at their overhead prey—insects, spiders, even small lizards.

Schuster has published his findings in major science journals, and afterward has endured his share of collegial skepticism. Now he is enjoying a slowly growing acceptance of his core idea: that animals with small, simple brains can make complex cognitive decisions. That is, they can be intelligent.

"Intelligent circuitry can be assembled in any brain; that's my big belief," Schuster said. We were sitting in his tidy, book-lined office at the University of Erlangen in 2008, where he did several of his archerfish studies. (He's since moved to Bayreuth.) "It's not limited to those animals with large brains and many neurons," he said. "If evolution requires it [this kind of intelligent circuitry], it will be assembled—even with a small number of neurons."

Fish might seem an unlikely animal for a researcher to use as an example of an organism capable of making intelligent, complex decisions, which is one reason that Schuster uses them. "People don't expect much from fish, but that's where they're wrong," he said. "Fish are capable of much more than people think."

Schuster spoke with the confidence of someone with long years of experience and close observation—someone who, as biologists like to say, "has a feeling for the organism." When I asked him when he first began to study fish, I wasn't surprised that he replied, "From the time I was three, I liked animals, but particularly those that live in water—tadpoles, water striders, dolphins, and especially fish. I don't know why I was so attracted to fish," he continued. "Maybe because of the mystery of their watery environment. My parents didn't like fish at all. But they saw my interest, and every weekend they took me to the zoo in Stuttgart, where there were many wonderful tropical species. That was where I first saw archerfish. Of course I always wanted to see them shoot something, but I never did. There was an information sign that said this behavior was very rare."

Schuster stands about six feet tall. He has dark hair, polished skin that glows, and a happy, energetic manner—even when discussing neural anatomy, or how ecology affects the evolution of intelligence. He radiated even more enthusiasm when talking about his love of fish. His face lit up with a boyish grin, and his hazel eyes sparkled. And as he recalled the zoo's sign, he arched his eyebrows high above his glasses and laughed out loud—because in 1998, when he finally got some archerfish of his own and released them in an aquarium, they began shooting at everything. "Really! Just immediately! It was fantastic."

Schuster was beginning his academic career as a young professor when he finally witnessed the archerfish's hunting behavior. At the time, he was starting the study that has defined his career: Can some animals with small brains and relatively few neurons make fast, flexible, and complex decisions—that is, are they capable of more than simple, hardwired responses? Or is an animal's intelligence limited by the number of neurons it has, as was commonly believed? He chose fish to answer his questions, in part because the brains of fish are somewhat simple anatomically, but also because he loved them and knew how to care for them. He'd owned many tropical fish species since he was fourteen, when his parents gave him his first aquarium. For his research, Schuster needed a particular type of fish—a species that would reliably perform "some 'intelligent' behavior at top speed." He first explored the talents of various species of electric fish that search for their prey by emitting electrical signals. Despite some interesting results, none of the electric fish provided exactly what he was looking for—and then Schuster got the archerfish.

We often think of scientists as rationally planning each and every step of an experiment, but Schuster emphasized that his widely praised archerfish studies owe a great deal to curiosity, fun, and serendipity.

"Really, it all happened because of a mistake," Schuster said, his eyebrows rising again, this time in wonder at the mystery of chance. "I didn't set out to study archerfish. I'd ordered a small aquarium for my home, but the supplier sent me a huge one by mistake, so I brought it to my lab. I bought the archerfish to put in this tank. They were just for fun." He considered them his "pets," not research subjects. "It actually took me an embarrassing amount of time—three years—before I recognized their potential for my studies," he said, amused by his own slow thinking.

By then, Schuster had watched the archerfish long enough to wonder about many of their abilities. No matter how difficult a target he gave them, they almost always managed to hit it. They willingly fired at insects as well as at geometric shapes, which suggested that their shots weren't hardwired; they had a degree of mental flexibility. And they did everything—spotting their prey, firing, and retrieving it—so speedily. "It was just the behavior I was looking for," he said.

Schuster and his students laid out a series of questions to answer: How did the archerfish compute their shots? They looked at their prey from underwater, which distorted the view, yet they accurately judged how to hit it from different angles and distances. How did they manage this feat? How did they determine when and where their prey would land in the water? The archerfish were masters at blasting stationary prey. Could they also knock down flying insects with the same precision?

And most challenging of all: Which cells in the archerfish's brain handle these many decisions? With his chosen species, Schuster thought he might be able to show decision making as it occurred, just as neurobiologists have identified the actual neurons that the invertebrate snail *Aplysia* employs when learning and forming long-term memories. Perhaps the archerfish could teach us general principles about how the brains of vertebrates operate at the cellular level.

Sometimes when researchers set out to investigate an animal's ability to solve a problem, the creature reveals something even more unexpected and surprising about its mental talents—something the scientists weren't even looking for. Schuster and his students had recently experienced such a serendipitous discovery. "Really, it wasn't something we planned," Schuster said, jumping up from his desk. "I'll explain all about it, but first, you must see the fish!"

SCHUSTER'S LAB was just a few doors away from his office, on the second floor of a modern concrete-and-glass multistory building, but it had the feel of a cluttered, if organized, garage. Metal shelves along one wall held a dozen fish tanks, lengths of rubber tubing, electrical cords, spare water filters and heaters, rolls of plastic tape, metal strips, and other mechanical oddments. In two small adjoining rooms there were more aquaria containing tropical fish and seahorses. The air was humid because of the tanks, and the lab sounded something like a health spa because of the bubbling filters. Thick, black electrical cables snaked across the floor, delivering power to all the tanks' filters and heaters—and I quickly learned to watch out for these, since one misstep would have sent me crashing into an aquarium. "Mind your step," the scientists said helpfully over and over again.

The archerfish caught my eye the moment I walked into the lab: eight silver-and-black dappled fish looking back at me from inside a wide, square tank. As big as the palm of my hand and shiny, they were shaped like Paleolithic arrowheads, broad and flat, with pointy heads. They weren't swimming in slow laps or zooming back and forth, as aquarium fish often do; they were lined up in a row at the front of the tank, their dime-sized, black eyes fixed on us. There was something so expectant in their pose and attentive gaze that I turned to Schuster and asked, "Are they waiting for you?" As soon as I said that, I blushed and thought to myself, "What a *stupid* question. Of course the fish aren't waiting for him." I expected to see him wince.

But Schuster was beaming. "Yes, yes. It looks that way," he said, stretching his neck toward the tank. I noticed that he didn't get too close. "They have these large eyes, and they do watch us as we move around the room."

"They shoot at us, too," said Mario Voss, one of Schuster's graduate students. "Like this." Voss stepped up to the tank, and leaned his head over the edge. The fish turned toward him, and in the next second, one blasted Voss in the eye with a sharp jet of water. "Oww!" cried Voss, jerking his head back and laughing as he wiped away the water. "A good shot."

Thomas Schlegel, who was completing his doctorate, didn't approach the tank. He'd been dodging water bullets for four years and was tired of being a target. "They shoot us in the eyes and nose all the time," he said. "It can be irritating after the thirtieth time in one day. But you should try it," he urged.

I stepped to the tank's edge, leaned in, and concentrated on keeping my eyes open. Which fish would be the shooter? The fish were all facing me, but one in particular seemed to be staring directly at my left eye, like a hunter targeting his prey. Wham! The water hit my pupil with such force that I jerked back, just as Voss had. It was like being shot by your kid brother's high-pressure water pistol. Laughing, I wiped my left eye but stayed by the tank, my left hand resting on its edge. Another fish quickly seized the opportunity to blast the diamond in my engagement ring, while a third targeted the red carnelian stone in my earring, and yet another shot my right eye.

"They are really good," I said, jerking back. "But do they really think my eyes and earrings are insects?"

"I think they can tell what is an eye and a nose, or earrings," Schuster said. "They don't really react like they expect this 'prey' to fall into the water. But eyes and noses do interest them."

"Maybe they just like our reaction," added Schlegel. "Most of the time, it must be boring for them, since they're just waiting for us to set up experiments. So shooting our eyes might be fun."

We stood there in a row, the four of us, looking at the row of archerfish, who had lined up again to look at us.

What were they thinking?

One of the archerfish waggled his head, stopped, and then did it again. "I could never prove it, but I always think when he does that he's trying to get me to feed him," said Schlegel, whose laptop sat opened on a desk beside the tank. He usually worked there, sitting with his back to the fish. "Sometimes I get the feeling they're watching me," he said. "And when I turn to look at them, there they are, lined up like this, waiting like dogs for their humans. They try to catch your attention."

"Most people don't look for learning or playful behavior or thinking or feelings in fish," said Schuster. "They think they're just, you know, *wet vegetables*. Carrots or cabbages with fins. Then they meet our archerfish. The fish look at them and shoot them in the eyes, and immediately people think, 'Well, these fish are intelligent.' And they *are* intelligent in some ways."

ARE FISH INTELLIGENT? Do they have thoughts—if only about how to manipulate their human caretakers? The notion is so new, or at least overlooked, that Schuster and others studying fish cognition are often on the defensive—

or worse, he says, completely ignored. Most cognition textbooks don't even mention fish, which led three fish experts, Kevin Laland, Culum Brown, and Jens Krause, in 2008 to produce the first one solely devoted to fish. In the 1950s, psychologists did show that mouthbreeding *Tilapia,* like rats and pigeons, willingly passed the standard stimulus-and-response learning test by pressing a lever (they struck it with their mouths) in exchange for a morsel of food. Despite that success, fish were still considered much like the dim-witted but charming Dory, the regal tang fish of Disney's *Finding Nemo*—who had a mere three-second memory. With that kind of limitation, the prevailing wisdom went, fish couldn't be expected to do very much, so it was also a given that they had no social skills, a minimal ability to learn, and very little sense of where they were or what they were doing from one moment to the next. They might be vertebrates like us, and they might hold claim to being the very first vertebrates (dating back to more than five hundred million years), but they were "lower vertebrates"—an old-fashioned label that put fish close to the bottom of the imaginary evolutionary ladder and kept them out of cognition labs throughout most of the twentieth century.

For much of the twentieth century, too, psychologists, rather than biologists, designed and ran the majority of cognitive tests on animals. They were most interested in finding out if animals could do some of the same things humans can do in terms of memory and learning—this was the original idea behind "comparative psychology."* As anyone who's ever taken a Psychology 101 course at college knows, the favorite research subjects were laboratory rats and pigeons, partly because they are easy to care for. Many species of fish aren't difficult to care for either. But because of their supposed limited memories, and because they are very distantly related to humans, fish didn't qualify for the psychologists' studies. After all, how could fish—or ants or any "lower animals"—shed light on the cognitive abilities of humans?

In recent years, however, as more biologists (and psychologists with an

* "There was actually very little that was 'comparative' about most comparative cognition labs in the past," one comparative animal psychologist told me. "Three animals were used: rats, pigeons, and college sophomores, preferably male. We laugh about it now. But the rats were the model for [understanding the cognition] of all mammals; pigeons were the models for all other animals; and the male sophomores represented all humans."

evolutionary perspective) have moved into animal cognition research, scientists have adopted a broader evolutionary approach—one that doesn't rank animals according to how close or distant they are from humans. Researchers seek instead to understand the ecological and social forces that may cause separate lineages of animals to evolve similar cognitive skills. Convergence in physical traits has long been accepted; for example, birds and bats evolved wings, not because they are close relatives, but because both had ancestors that began to pursue prey in the air. Scientists have only recently begun to apply the idea of convergence to cognition, but it is already helping researchers better understand the emergence of many mental skills such as social learning, teaching, planning for the future, and inventing tools. If animals—even those whose lineages parted ways long ago—face similar cognitive demands, they are apt to evolve similar cognitive abilities. The convergent approach has made the notion of a hierarchy of intellectual abilities passé. There aren't "lower" or "higher" species; nor can researchers simply wave away the cognitive abilities of an entire group of animals—such as all thirty thousand species of fish—by assuming that, on the basis of a few faulty experiments, they lack any.

In fact, fish—in spite of having the smallest brains relative to body size of all vertebrates—are capable of many intelligent behaviors, as a growing body of research shows. Studies of sticklebacks, guppies, zebrafish, and others have demonstrated that fish are equipped with long-term memories. They have personalities—some are shy and retiring, others are bold and aggressive—and problem-solving skills.* Their memories allow them to create and update mental maps of their highly variable aquatic worlds and to track social

* From my research, it seems that the first two papers that focused on animal personality were published in 1993—one was about personality in fish, the other about personality in octopuses. These days, it is increasingly accepted among biologists that all animals—from insects, fish, and octopuses to hyenas and gorillas—have personalities; that is, they display consistent individual differences in behavior. There are even animal personality scientists who are investigating how genes, gender, life experiences, and environmental conditions shape an animal's personality. In general, some animals are shy "sitters," sticking to the sidelines like wallflowers, while others are more daring "rovers," boldly heading into the unknown without any sign of fear. And there are variations in between these two extremes. Natural selection maintains the diversity of personality types because in some situations it is better to be retiring and cautious while in others it can pay to be adventurous and bold.

relationships. They can be as socially calculating as chimpanzees or humans, manipulating, punishing, deceiving, and befriending their fellows to get what they want. Siamese fighting fish remember the male who lost the last fight and treat him like a loser. Smart male Atlantic mollies that are busy courting a plump, fertile female will, if a rival shows up, turn their attention to a skinnier lady—since males often try to woo the same gal. Archerfish use water as a tool; mackerels hunt in groups, herding their prey. And two species of groupers hunt cooperatively with giant moray eels, an entirely different species; the groupers lead eels to prey hidden in rocks, then let the skinny predators flush out the fish—providing meals for both the groupers and eels. The two collaborate almost as successfully as humans and hunting dogs.

Many species of fish are also adept social learners, meaning that they learn by watching one another and that what they learn affects their interactions with others. Social learning is widespread among animals. For instance, male white-crowned sparrows acquire their songs by listening to other males, and chimpanzees use the method to learn how to crack open nuts with stones. Through social learning, fish gain knowledge about where to forage, which predators to avoid, and which mates to pursue. Darwin called this type of learning "imitation," but researchers today prefer the term *copying*. True imitation, as defined by cognitive scientists, requires the observer to repeat precisely the action that the demonstrator has performed—something that most animals rarely do. Further, the action must be novel, something the observer has never seen until the moment when he witnesses someone else doing it. Think of Meryl Streep. She can watch a person and listen to her talk for a few minutes, then instantly reproduce her accent and mannerisms. Aside from humans, only dolphins, elephants, bats, parrots, seals, and crows can do this type of imitating.

So when Schuster and his students found that their archerfish learned how to shoot at difficult and novel targets by watching another skilled fish perform the task, they knew they had a potential problem on their hands. It looked as if the archerfish were imitating.

"No one would imagine that a fish could do such a thing," Schuster said. "It wasn't something we were looking for, and we don't know what neural mechanism they use."

Actually, scientists don't know what neural mechanism we humans use to imitate another person's actions, either, or to do any type of social learning. Neuroscientists have suggested that we—as well as some primates and dolphins—may be using our mirror neuron cells, a network of specialized cells in the cortex, when we imitate or copy. They've watched the brains of rhesus monkeys (while wearing electrode caps) when the monkeys are doing something, such as grasping an object, and when the monkeys watch another perform the same action. In both cases—watching the task or doing it themselves—the monkeys' mirror cells fire. But because monkeys are not the perfect imitators that we are, mirror neuron cells don't explain the entire process. It may be that these neurons serve a more general purpose for social learning overall, some scientists think.

No scientist has reported discovering mirror neuron cells in the brains of fish; so far, they've been found only in mammals.

There's a long-standing rule in animal cognition studies called Morgan's Canon. The rule is named for the British psychologist C. Lloyd Morgan, who proposed it in the late nineteenth century as a way of reining in some of the wilder anthropomorphic claims about animals' cognitive abilities. Considering its influence, the rule is rather a disappointment to read (it's not simple or elegantly phrased), but here it is: "In no case is an animal activity to be interpreted in terms of higher psychological processes, if it can be fairly interpreted in terms of processes which stand lower in the scale of psychological evolution and development." In other words, scientists should look for the simplest explanation for an animal's behavior.

So even though the archerfishes' actions looked remarkably like the textbook definition of imitating, Schuster abided by Morgan's Canon and did not use this hot-button word. He and his coauthors limited themselves to saying the fish were "copying" the skilled performer.

But Schlegel did say, "Really this is incredible." And Schuster concurred, "We had not expected it."

SCHUSTER'S SURPRISE DISCOVERY grew out of a series of experiments with the archerfish over a seven-year period. He designed each of the experiments to test his larger theory that small-brained animals are not simply hardwired but have some capacity to form flexible decisions. To build a case, he set out to tease apart the archerfish's decision-making process as it was firing its quick, accurate shots.

The old adage "Haste makes waste" applies to animals, too. Scientists have shown that creatures that make speedy decisions often make mistakes. But not the archerfish. They so rarely miss their target that they've become an Internet phenomenon with their own YouTube videos. Prior to Schuster's studies, it was widely believed that the archerfish always managed to hit their prey because the fish were completely hardwired; there was nothing cognitive—that is, no learning or decision making—required. Schuster and his students have shown that this is not the case. The archerfish learn how to make all of their shots, and in order to hit a target they must make many calculations and decisions, some in mere milliseconds.

To show me exactly how an archerfish hits its prey with a shot of water, Schuster and I returned to his office, where we watched a slow-motion video on his computer. On the screen, an archerfish came into view, sculling and surveying a leafy twig hanging over the water, about three feet away. A beetle rested on the twig. "Now watch what he does," Schuster said. The archerfish barely poked his snout above the surface and fired a stream of water, scoring a direct hit. "The archerfish has a groove or slot on the roof of his mouth, and a very oddly shaped tongue," Schuster explained. "With those, he forms a tube, like a blowpipe or gun barrel." The fish forces water down the tube by snapping shut his (or her) gills. "They're very fast and accurate, and they know just how much force they must use to knock down their prey." On the screen, archerfish blasted spiders, butterflies, grasshoppers, and a row of ants one by one.

The fish were like video arcade hotshots. They were astonishingly fast and accurate, which made them fun to watch. But why was Schuster so certain their skills weren't simply reflexive behaviors?

Schuster smiled at my question and, with a kind of "just-watch-this" expression, opened another video. It showed an archerfish shooting a spider in slow motion. The fish's spitted stream coursed through the air like a wet projectile, knocking the spider into the pool below. "To make that shot, the fish must calculate everything from the distance of the target, to the amount of force he needs to use to dislodge it, to where it will land, and how fast he must accelerate to grab it," Schuster explained. Because young archerfish are poor shots, "they must learn how to do most of this," he added. "That means it is cognitive; it is not because of some built-in, hardwired neuronal machinery."

In the lab, Schuster and his students videotape the fish firing at various targets, stationary, moving, horizontal, vertical, large, and small. Then the researchers analyze the high-speed images frame by frame. From these experiments, the scientists know that most of the fish can manage a horizontal shot but must practice to become proficient at high, vertical ones. Similarly, most archerfish are very good at stationary targets, but even the best sharpshooters can't hit moving prey until they've worked at this skill for days, taking hundreds of practice shots. Every increase in target height or speed momentarily sets the fish back, forcing them "to adjust their shots and make new calculations," Schuster said.

Archerfish have "general rules" that underlie their calculations, Schuster said. Rather than simply firing an all-or-nothing blast of water, the fish adjust the stream—not for the size of the prey but for the amount of force an animal uses to stick to a surface. Other researchers have shown that this adhesive force is closely proportional to the animal's size. "It is that adhesive force the archerfish actually target," said Schuster. "They hit their prey with a blast about ten times stronger" than the force the animal uses to stay stuck to a branch.

The most demanding—and cognitively challenging—task for the archerfish comes after it has hit its prey. At that point, the fish must figure out where the prey will land and how fast it must swim to grab it—calculations the fish must also learn how to make. The fish doesn't do this by tracking the arc of its prey (which is what baseball players do when figuring out where to catch a fly ball). Tracking would waste time. Instead, Schuster and Schlegel discovered that as soon as the prey begins to tumble, the archerfish determines where it will land and how quickly the archerfish must accelerate to get there to grab the prey when it touches the water. The shooter cannot dawdle—in the wild, the archerfish live in schools and must compete with one another and numerous other larger fish for every insect they shoot down. They have to think in the blink of an eye.

"Archerfish make those calculations in forty milliseconds [forty thousandths of a second]—a fraction of a second," said Schuster. "Now, some people criticized us and said this can't involve cognition because it is too fast. So if the fish did it more slowly, then it would be cognitive?" Schuster laughed and shook his head. "We make the same types of very fast decisions. No one would say it isn't cognitive."

Actually, the archerfish can retrieve prey more slowly—and do so, when it is the only fish in the tank. After all, why waste the energy if there's not a race?

It doesn't matter that archerfish are making their decisions and calculations using piscine brain cells, Schuster stressed. From comparative anatomical studies, neuroscientists know that the brains of all vertebrates—fish as well as humans—operate in the same way, transmitting signals between cells chemically and electrically.

For fish, some of the most important neural signals concern predators. The mere glimpse of a predator triggers a six-neuron circuit in a fish's hindbrain, causing the fish to turn away instantly. Archerfish make this move (called a C-start, after the shape a fish's body assumes as it darts away) when they flee a predator. They also make a C-start when they race to retrieve their prey—which means they're using the same neurons for both tasks, even if one (prey retrieval) is more complex than the other (escape). Evolution often works this way, Schuster says, adapting an existing mechanism to a new use. The main difference between the two types of C-starts is that the one connected to prey retrieval is triggered by the motion of the falling prey.

"The fish gets a very brief glimpse [of its prey], and from that small amount of information, using this small neural network, it makes all these calculations," Schuster said. "It shows that accurate and complex decisions don't require large brains. The job can be done with a relatively small number of neurons. The archerfish give us the chance to understand at the cellular level how brains in general, large or small, make decisions"—one of the holy grails of neurological research.

WE WENT BACK to the lab for another round of archerfish target practice. Schlegel had hung a 4-inch-diameter clear plastic disk over the tank. It floated midair about one foot above the water's surface. From a box of frozen flies, Schlegel selected one and quickly dunked it in the tank to moisten it. Then, cupping his hands together so that the fish couldn't see the fly, he stuck it on the bottom of the disk.

"That's my sign to them; cupping my hands like this," he said. "They know, 'Okay, I need to be attentive.'"

And like good pupils, the eight archerfish quickly lined up at the other end of the aquarium to face the disk. As soon as Schlegel dropped his hands and stepped back, zap!, one of the fish blasted the fly into the water. All the fish raced to retrieve the prize—and it looked to me that a fish other than the shooter won the race.

"Yes," Schlegel said, pointing to the one I had managed to keep track of. "He got it and he wasn't the shooter."

Schlegel attached another fly to the disk, and whap!, down it dropped. This time, the fish had such a splashy race that I lost sight of the winner. But Schlegel said, confidently, "That time, the shooter got it."

Because Schlegel hadn't marked the fish, as Nigel Franks had his ants, it wasn't clear to me what method he was using to identify them. It seemed very difficult in a rough race like this last one to really know which fish had done what. The distinctive clues were there, though, Schlegel said, in the archerfishes' behaviors. "That's what we use to tell them apart," he said. "After you've worked with them for a few weeks, you can really see that they are different." Some never shoot when they're in a school; they shoot only when they're alone in a tank and then fire away willingly. "They're like cheaters," said Schlegel. "They let the others do the work but then they try to get the fly."

The more dominant fish chase the others away, trying to prevent them from shooting at all, and defend the spot they prefer to shoot from.

The differences in the archerfishes' behaviors had not posed any real problems until Schuster (and other students no longer at the lab) began the moving-target study with five particularly good sharpshooters. At least, they were good sharpshooters until the target began to move. Then they were "terrible," unable to hit flies traveling even at the pace of a snail.

Slowly, over several days, "they got better," Schlegel said, who wasn't at the lab during this study but who found the results so fascinating he had practically memorized the paper. He paused to correct himself. "Well, *one* got better." This supersharpshooter was hitting targets that moved as fast as a hornet. But he was also an aquarium bully, hogging all the shots and nudging the others away so that they couldn't shoot.

"Finally we just removed him," said Schuster. "We thought he was spoiling our experiment. Now we were going to have to spend more days training these other fish."

Instead, all four were immediately almost as good as the bully at shooting the moving targets.

"They shot everything," said Schlegel, from the snail-slow flies to those speeding by like a bullet. "They had learned everything" about target speed and the effects of gravity "by watching this other fish, the top sharpshooter. And that means they had taken the viewpoint of the other fish."

Somehow, the observing fish were able to essentially imagine or mentally picture themselves in the position of the shooting fish. It was as if you or I had watched Kobe Bryant make a tough layup shot one hundred times. Then we got up from the bench, took the ball, and made the shot ourselves—not once, but every time thereafter. Anyone who accomplished such a feat would be signed up by a rival team.

To make certain that they had not somehow given the observing fish inadvertent clues, Schuster and his students also tested a control group of four other fish that did not have a dominant fish. The scientists let the fish watch the moving target but nudged them away if they moved into the firing position (just as the bully in the original group had). Later, when the researchers let these four archerfish fire at the targets, they performed exactly as most spectators do—badly. They required more than twenty attempts to score. "It showed that the first four fish had learned their skills by watching the bully," Schuster said.

As with other aspects of copying or imitating, scientists do not know how humans, let alone fish, manage to take another individual's viewpoint. So far, there is no known process that explains how any animal's brain—simple or complex—accomplishes this feat.

Did the archerfish copy or imitate? Schuster and Schlegel smiled and shrugged. They weren't going to be entangled in a debate about definitions. What mattered was that once again they'd shown that what the archerfish do involves cognition.

FISH ARE ONE of the few wild animals that most of us consume without any feelings of guilt. I love a good salmon steak, or filet of freshly caught trout, and I rarely think about how that fish ended up in my freezer or what it felt as it was pulled from the water. I'm more worried, as I suspect many others are, about the numbers of fish that are removed from the oceans and rivers each year through industrial fishing. The numbers are staggering—more than eighty million tons, or one trillion individual fish, are caught and killed annually. Of these, more than one-third are converted to fish oil and fish meal for use in animal feed and aquaculture (ironically, for feeding farm-raised fish). Conservationists and ecologists warn us that we may be causing the extinction of several large marine species, such as Atlantic bluefin tuna, swordfish, halibut, and sharks, and altering aquatic environments beyond repair. Equally alarming are recent reports that we're already fully using or overfishing some 75 percent of the world's fisheries. The United Nation's Food and Agriculture Organization and other agencies predict that we will run out of fish to eat unless we develop more sustainable ways to go about our fishing.

What we generally don't fret about is the act of fishing itself. Until recently, scientists didn't either. They didn't study such things as whether industrial fishing is humane, or investigate fishing from the fish's point of view. But studies such as Schuster's and others into the cognitive abilities of fish may change consumers' attitudes—which, in turn, might lead to changes in the fishing industry.* We've been raised to believe that fish do not feel pain. Yet if you knew that fish *could* suffer pain the way cows or lambs or chickens do on factory farms, would you buy cod that have been hooked and dragged on long lines, or tuna that were captured in purse-seine nets before being hauled to the surface, where they die—gasping and thrashing—of suffocation?

It's not that you have to be smart to be mentally aware that you hurt. Honeybees, ants, and many invertebrates can perform many cognitively demanding tasks. Researchers don't, however, interpret the complex dance of the honeybee or the real-estate appraisals of the rock ants as indications that bees and ants have emotional lives—they are not sentient and do not experience feelings and emotions. Nor does a fish's ability to make quick, flexible decisions necessarily mean that it suffers when pulled from the sea. But animals with more mental flexibility typically possess more complex sensory and nervous systems, suggesting that they are the most likely to have the neural hardware that underlies suffering.

Although scientists don't understand fully the relationship between cognition and sentience, they know that the two are intertwined—that one ability informs the other. For instance, human athletes sometimes compete even when in great pain; they are aware of the pain, but they choose to ignore it to achieve another goal. Rats also can elect to ignore a painful heat sensor attached to their tails if they smell a cat—it's better for the rat to remain in a frozen position on the hot plate, as it were, than move and expose itself to a predator. Can fish do something like this? And if they are able to experience the sensation of pain, are they mentally conscious of suffering?

For Victoria Braithwaite, a fish biologist at Pennsylvania State University, the question about whether fish feel pain came up as she investigated

* Changing industrial fishing methods may not be as far-fetched as it sounds. The whaling industry was almost completely transformed when the International Whaling Commission adopted a moratorium in 1982 on commercial whaling, while allowing quotas for aboriginal subsistence hunting. Although a few nations, notably Japan, Norway, Iceland, and the Russian Federation, oppose the moratorium and continue to whale, most countries abide by the rule. The IWC also encourages more humane killing operations.

methods to increase the survival of salmon and trout raised in hatcheries. (Many hatchery-raised fish are returned to the wild in hopes of restoring dwindling populations. But the success rate for this practice is low—in fact, fewer than 1 percent of the fish survive more than a few weeks after being released.) Braithwaite found that hatchery fish were often kept in crowded conditions; many had lesions on their sides from scraping against the walls, and some fish attacked others, biting their fins. Braithwaite wondered why the aggressive fish targeted the others' fins in particular. Are the fins of fish especially sensitive to pain?

Braithwaite was based at the University of Edinburgh when she started this research in 1999, so she and her colleague Mike Gentle applied for a British government grant. The grants committee agreed that their question was worth pursuing but suggested a slight change.

"They wanted to know, do fish feel pain when they're *hooked*," Braithwaite said. "They went right to the heart of the matter."

The committee's question, of course, also altered the emphasis. A fish usually encounters a hook that's dangling from an angler's rod, so the committee might as well have asked, "Does fishing cause a fish to suffer pain?" It was an awkward question to ask in Britain, where sport fishing is a cherished hobby, but Braithwaite and her research group tackled it head on.

Round faced, blond, and blue-eyed, Braithwaite has a rugged, practical air. She's the mother of two teenage sons and a fishing advocate. Although she isn't an active angler these days, she has fished in the past, and she enjoys eating fish almost as much as she enjoys watching them in the wild. Fishing provides a good way, she thinks, to help children develop a love of nature. She's studied the health and cognition of fish for nearly twenty years and has grown increasingly worried about intensive fish farming; this was what really spurred her research. She thinks that the conditions on most fish farms probably cause the fish to suffer, which adversely affects their growth, as is often the case with stressed animals. She would like to change how fish are farmed so that their well-being is considered.

"We're concerned about the welfare of chickens, pigs, and cows on farms. Why not fish? They are animals, too," she says.

When she and her colleagues looked into the fish-farming industry, they were stunned to learn that there were really no rules or any form of guidance about how best to raise and care for the fish. Since the 1980s, the animal welfare movement had led to many positive changes for animals farmed on land, but no one seemed to consider what life on a fish farm might be like for the fish. Braithwaite was surprised to discover how little was known about pain in fish. Here it was, the end of the twentieth century, yet she could find only several cursory studies that basically concluded that fish are something like ants—they might have a reflexive response to pain, but they don't process it mentally. No scientist had actually studied a fish's anatomy closely enough to determine if fish have the type of nerve cells and fibers that other vertebrates, such as birds and mammals, possess for sending a signal to the brain that something hurts.

Pain had been missed in fish, Braithwaite thinks, because we humans really do not understand them. Their faces are immobile and don't convey the type of emotional signals we're used to reading. We also have trouble understanding and appreciating what life is like in their aquatic world. Besides their ability to think, fish have a surprising number of ways to gather information about their environment. Many can glean scents carried in water, and most have color vision and acute hearing. Fish make a range of calls—squeaks, squeals, chirps, barks, moans, groans, and hums—and use these to find mates, warn others to stay away, or alert their school to danger. Like birds, many build nests to raise their young and sing to woo their partners. Numerous species of fish communicate with such low-frequency sounds that we cannot hear them without special instruments. We mostly miss the calls of fish that are within our hearing range, too, simply because we don't hear very well underwater. If you've ever snorkeled among a school of fishes on a coral reef, you've no doubt been dazzled by their flashy colors and curious shapes—but you probably did not hear their vocalizations or any of the other natural noises emanating from their coral reef home. Marine biologists using hydrophones have found that reefs crackle with the electric pops of snapping shrimp, the clatter of crabs, and the songs of fish. The sounds of each reef are apparently so unique that the young of some fish can identify where they were hatched and use these noises to find their way home again, although ocean currents had carried the fish miles away when they were larvae.

Wouldn't it make sense for such sensory-laden creatures to also have some ability to perceive pain?

This was Braithwaite's contention. All animals, even jellyfish, can sense disagreeable things. Jellyfish will swim away from a place in an aquarium where they've received a shock, but they don't have a brain to link that unpleasant sensation to an emotion. It's unlikely that they suffer in the way that a cat or bird does from something painful. The ability to detect something that hurts your flesh is called nociception, and the nerve cells that give you this ability are called nociceptors. Nociceptors are found in birds, mammals, and amphibians, and even in invertebrates, such as leeches and sea slugs. Many of the classic anatomical studies on how nociceptors work were done on leeches and sea slugs. If you prod a sea slug with a sharp object, it withdraws in response to the pain. Scientists don't think that the sea slug is experiencing the pain mentally, because they have not found the specialized nerve fibers that are necessary to carry the signal of pain to its brain.*

Braithwaite and her colleagues in Scotland began looking for nociceptors and the specialized nerve fibers in rainbow trout in 2001 and published their first major article on the subject in 2003. I had met her at her campus office, and as she recited the early history of her group's studies she called up this key paper on her computer and clicked on an image that best captured their findings. It was an enlarged color photograph of the face and head of a rainbow trout. She had placed small colored triangles, diamonds, and hexagons along the fish's upper and lower lips, face, and chin, near its gills, and around its eyes—each pinpoint shape designating one of three types of nociceptors. "Those are just the pain receptor cells on the trout's face and head," she said. "We found twenty-two of them." The scientists also found the requisite nerve fibers for moving the cells' signals to the brain—proving that fish do have the physical and neural hardware to detect and process pain.

"There are a lot on their lips," I commented, trying to sound objective,

* The question of pain in invertebrates is far from settled. Recent studies indicate that prawns and hermit crabs retain memories of painful events. Octopuses and squids are also sufficiently intelligent—able to figure out how to open lids on jars or escape from their enclosures—that scientists studying them wonder if cephalopods' nervous systems may be more developed than we realize. They may even experience something akin to emotions.

although her diagram of "places to pierce on a fish if you want to cause it pain" had sent my objectivity right out the door. I had grown up in Southern California and had spent many summers hiking in the nearby San Gabriel Mountains, as well as the Sierra Nevada range. My parents loved to camp and fish and taught me to do and love both. As a special present on my eighth birthday, they'd given me my very own rod and reel, and I'd hooked fish (mostly trout) at each one of the pain receptor points Braithwaite was now describing. I was what my husband and I jokingly called a "heavy-metal fisher," since I used fishing lures that sparkled like a minnow. They were also usually armed with three sharp, barbed hooks—a design meant to catch and hold, even if the fish decided after taking a bite that it didn't really want this particular meal. Sometimes the hooks seemed to latch onto fish that might have just swum up to look at the lure. I'd hauled in trout that were impaled in their eyes, gills, cheeks, and chins. Had I really believed the fish didn't feel something when I pulled them, flopping and wriggling—and yes, gasping for air—from the water? Who convinced me of that? I tried not to look at the photo and turned my attention to Braithwaite, who was eagerly explaining the details of each type of pain receptor. "So," she concluded, "I was able to tell the granting committee, 'Yes, hooks hurt fish.'"

"Great," I said.

"But, you know of course, that's not really what they were asking. What the committee and most of us really want to know is: Do fish *suffer* when they sense that pain? Suffering is the cognitive side of pain. It's what we humans worry about; we don't like to cause unnecessary suffering," she said.

"No, of course not," I agreed. Silently, I wondered if I would ever be able to go fishing again. "No, of course not," the voice in my brain said.

Braithwaite clicked on the next image, an illustration of a fish's brain. It looked like a bundle of small potatoes. She pointed out various features and explained that the brains of fish had long been regarded as missing many of the key structures that are found in the brains of mammals. Fish were thought to not have an amygdala, for instance, an area that helps process basic emotions, such as fear, and positive and negative rewards. Just a few years ago, however, an anatomist discovered that fish do indeed have a similar structure in their brains. (It had been missed because fishes' brains grow outward, rather than inward as ours do; all their special structures are on the outside, not the inside of their brains.) Braithwaite pointed to it on the screen. A fish's amygdala (technically called the medial pallial nucleus) is part of its forebrain, one of the larger, potato-like structures.

"Now, this discovery [of the fish's amygdala] changes everything," Braithwaite said, because similar cognitive structures have been found to work similarly in all vertebrates. It's why rats are often used as human surrogates to test psychobiological drugs, for instance. "It means we can get at fishes' inner states, their feelings," she continued. "What gives them a positive feeling? What puts them in a negative or depressed state?"

I imagined that a hook through the lip probably would do the latter. "Yes," she said. "But we need to take that one step further and see if they have a 'mental representation' of this pain."

In Scotland, she and her students had shown that trout given an injection in their lips of a tiny amount of bee venom or acetic acid behaved in a way that suggested "discomfort." The trout rocked back and forth, something that primates do when they are distressed. Those injected with acid rubbed their lips on the gravel and against the sides of the tank, much as we do when we try to soothe a spot that hurts by rubbing it. No one had ever reported these behaviors in fish before. They weren't just simple, reflexive behaviors either. Tellingly, for three hours afterward, the injected fish didn't touch a morsel of food, although fish that had been injected with saline—and so felt the prick of the needle—ate with as much gusto as did a group of untreated fish.

"I think they were responding to something painful," Braithwaite said. "But we have to be very careful in our wording in our papers, because people accuse us of saying that fish feel pain like we do. And we're not saying that."

Since emotions cannot be measured directly in animals (at least no one has yet figured out how to do this), Braithwaite must use an indirect measure—which is the fish's behavior after a painful experience. At the time of my visit, she was beginning a new round of experiments and had designed these to take advantage of fishes' lateralized brains. Like us—and all vertebrates, for that matter—fish have brains that are divided into two hemispheres. Our lateralized brains make us either left- or right-handed, among other traits. Similarly, fish can use their left or right eyes for two different tasks, keeping an eye out for predators with one while looking for food with the other.

Braithwaite's lab is housed in two adjacent rooms in the basement of her office building, and she led the way down the stairs. Much like Schuster's lab, the rooms were crammed with bubbling aquaria and assorted supplies and equipment. Braithwaite had only recently moved to Penn State and had not yet received the stickleback fish she wanted for her new experiments. But she was doing some similar work with guppies and bishops, which are closely related species, and we stopped at one tank to admire the fishes' bright colors and darting moves.

As a demonstration, one of her students had set up a tank with a T-shaped maze made of clear plastic dividers. Braithwaite scooped a male guppy from another aquarium using a cup ("it doesn't stress the fish as much as a net does"), and placed the fish in the experimental tank. The guppy was an old hand at maze tests and swam down the corridor. At the end, he came to a glass barrier and the T-junction where he could turn left or right. Behind the barrier, stuck in the sand, was something he had never seen before: a small, yellow, plastic cross. Fish in a "positive frame of mind" realize that the cross is new and therefore potentially dangerous. They approach the barrier cautiously, as this male was doing. "See that," said Braithwaite. "He's alert; he's keeping a careful eye out for that cross." The guppy turned left and swam parallel to the barrier, so that his right eye tracked the cross. Then, zoom, he was out of the maze and ready for his reward. Braithwaite dropped a treat in the water for him.

"That's the way a fish should behave if it sees something new or a predator," Braithwaite said. "It's the smart and safe thing to do."

But what if a fish is first given a small prick of bee venom in its lip? How will it respond? Will it be able to register the danger? Or will it be distracted because of the pain? She would be investigating these questions with the sticklebacks in large, outdoor mazes she planned to build. But for the moment, she said, her previous studies with trout suggest that fish in pain will not behave normally, that they will either be less alert to the potential danger or ignore it completely. "I think it shows that painful experiences do affect the ability of fish to make decisions; that they're in a vulnerable state after something painful happens to them because they are suffering. Fish have the cognitive capacity to experience emotions, and are self-aware, and conscious," Braithwaite said.

Braithwaite's conclusions were widely publicized. Anglers were understandably distressed but "also willing to think about ways to limit the harm to fish," Braithwaite said. Others were not so accepting. Certain researchers continued to argue that the trout were only reacting reflexively, while others, especially human psychologists, refused to accept her evidence that fish are self-aware and conscious. They insisted that because pain is a psychological experience, fish cannot suffer.

Several weeks before my visit, Braithwaite had spoken to a group of human psychologists who wondered if she would next be asking if insects suffer emotionally.

"My response was: Well, shouldn't we have a look? Wouldn't it be interesting to know if insects have some kind of mental representation of pain? And if they don't, why not? It seems very unlikely that insects would have feelings, yet these kinds of emotions have clearly shaped us and other vertebrates. They help us learn to protect ourselves—to say, 'Oh, that *really* hurts; I *really* don't want to face that again.' What's wrong with exploring this question in other animals?"

Darwin would have said, "Nothing." He would expect to find some degree of our own senses of pain and suffering in other animals, from primates to insects. For him, all creatures were capable of intense emotions. Even insects could express "anger, terror, jealousy, and love by their stridulations [the sounds crickets, for instance, make by rubbing their back legs together]," he wrote, and made other, grating noises "from distress or fear." He wrote descriptively of the pain animals felt, and their suffering; and he noted that if they suffered unduly, they became dispirited, depressed, and lethargic—just like Braithwaite's venom-injected fish.

Yet for much of the last century scientists largely regarded cognition and emotion as separate entities. Since the 1990s, however, that idea has been discarded, and these days the two are considered as inextricably linked—at least in us and our close primate kin. What if, as Darwin argued, this is true for many other types of creatures—even those that seem most driven by instinct, such as fish and insects? In untangling the evolutionary roots of cognition, researchers are inevitably drawn into the origins of our emotions as well.

Before leaving Braithwaite's lab, I told her I felt haunted by the diagram of the trout's pain receptors. I wasn't sure I would enjoy fishing again. "Use barbless hooks," she laughed, "particularly if you're going to release your catch."

She wasn't sure if catching and releasing trout was always the best idea, since anglers and bioethicists are still debating if this truly benefits the fish.* But any severely injured fish should be dispatched quickly and not left to suffer.

Fish are unable, of course, to speak for themselves about how we treat them. They have ways of communicating with each other and, in some cases, even with other species of fish, as the groupers and eels do. As with most animals, their inability to communicate directly with us puts them at a disadvantage. They cannot argue for their rights or how they might best be treated or farmed or managed in the wild. Most animals have no voice that we can hear, unless we speak up for them.

And even if an animal could talk, would we listen?

---

* In a 2007 study, researchers at the University of Illinois showed that catch-and-release works best if the fish are kept out of water for only four minutes or less. Longer than that and their physiological condition deteriorates, making it more likely that they will be caught by a predator when returned to the water. Other studies have shown that fish have a better chance at survival if anglers use knotted nylon or rubber nets and wet their hands before touching the fish. And angler should always remove the hook from the mouth of any fish they release.

3

BIRDS WITH BRAINS

In the very earliest times, when both people and animals lived on earth, a person could become an animal if he wanted to, and an animal could become a human being. Sometimes they were people and sometimes animals and there was no difference. All spoke the same language. That was the time when words were like magic.

NALUNGLAQ, A NETSILIK ESKIMO

Nearly all the scientists I met for this book wished they could talk to their animals. "Of course, it would be much easier if I could ask them these questions," they would say. "But since that's not possible, I have to find another way" to get inside their minds. Many of them had dreamed that they talked to their animal and that he or she talked back. They envied Doctor Doolittle.

But what if you *could* ask your animal questions—just speak to him or her directly, and get a reply? Would we better understand the minds of our fellow creatures?

Irene Pepperberg used this approach with Alex, her male African gray parrot, whom she called her "close colleague." She used those words so that other people (primarily her human colleagues) would understand that Alex wasn't her pet and that she wasn't emotionally attached to him. She wasn't Alex's soccer mom, rooting for him to pass the cognitive challenges she gave him. She wanted to understand objectively how the minds of birds—specifically, parrots—worked.

I met Alex and Pepperberg in her lab at Brandeis University in Massachusetts, where she is an adjunct professor. Sadly, Alex has since died from a heart arrhythmia, but at that time he and Irene were in the twenty-ninth year of what Pepperberg called their Avian Learning Experiment. (Alex's name was an acronym derived from that title.) They had come closer than any human-animal pair ever has to achieving the seemingly impossible dream of verbal interspecies communication.

"I thought if Alex learned to communicate, I could ask him questions about how parrots see the world—how they reason. It would be a new way to understand avian cognition," Pepperberg told me.

Pepperberg and Alex were seated—she at a small desk, he on top of his cage—in her lab. She was stylishly dressed in a gray miniskirt and rose-colored sweater, colors that suited her complexion and dark hair and eyes. Gray and red are also the two main hues in the plumage of African gray parrots. When Pepperberg stood next to Alex, they looked like a human-bird version of a mother-toddler pair dressed in matching sweater sets. Pepperberg said she chose the colors not because they were those of her parrots but because they looked good on her. She wasn't a crazy bird lady. Nevertheless, there was something charming about the two being "dressed alike"—something that suggested a warmer relationship than the words "close colleague" allowed.

It was just after 8:30 a.m., and Pepperberg's two assistants were standing at a sink about three steps away from her desk, chopping apples and bananas for Alex and two other African gray parrots, Griffin and Arthur. They were perched in the back of the room, which was small and narrow, across from Alex. The lab's walls were covered with beige, sound-deadening curtains, making the room seem all the more cramped. Pepperberg had squeezed in a chair for me next to hers. She was simultaneously sorting through her mail, introducing me to everyone, and explaining the parrots' routine. The staff had already tidied their cages and spread clean newspapers on the floor, and the parrots were now waiting to be fed.

The humans and the two younger parrots served as Alex's flock, Pepperberg explained, providing the social input all parrots crave. Alex was the oldest bird, and, I quickly deduced, everyone (human and parrot) catered to him. Like any flock, this one—as small as it was—had its dramas. Alex dominated his fellow parrots, acted huffy at times around Pepperberg, tolerated the other female humans, and fell to pieces over a male assistant who stopped by for a visit. For some unknown reason, when it came to what might be called social affairs, Alex always preferred men, especially tall, blond men, like this assistant. "If you were a man," Pepperberg said, after noting Alex's aloofness toward me, "he'd be on your shoulder in a second, barfing cashews in your ear—that's the parrot equivalent of love."

As it was, Alex gave me a quick glance, decided I wasn't his type, and ignored me. So, like a groupie, I could study him—the world's most famous bird!—in great detail from just a few feet away. Alex was about a foot long, from head to tail, and each of his dove gray feathers was trimmed with white as if edged with lace. White feathers framed his yellow eyes and set off his black, scimitar-shaped beak. Usually he had long, bright rose red tail feathers, too, Pepperberg said, but he was molting, and his "gorgeous tail" was but a stub. The two other parrots looked very much like him, although Griffin was bigger, but neither bird had Alex's confidence. They were wary of me and trembled if I looked their way, so I took my cue from Alex and ignored them.

"Parrots are slightly neophobic, afraid of new things and people," Pepperberg said, explaining Griffin's and Arthur's shivering feathers.

After Griffin and Arthur settled down, Pepperberg explained the parrots' lessons. All the birds were learning to emit what Pepperberg called "English labels" (that is, the sounds for objects) via a method she termed the "model/rival technique." Without going into all the details, here's a brief example of how this works: a parrot would be placed on a perch between two people. One of the people would hold up an item, such as a piece of wood, and ask the second person, "What's this? "*Wo-od*," the second person would reply, drawing the sound out in an enthusiastic singsong manner. The first person would then give the piece of wood to the second as a reward. This sequence would be repeated several times, with the humans taking turns playing the two roles, so that the parrot would learn from either one. Eventually, one person would ask the parrot the question. When the parrot replied, even initially with a poorly articulated version of the word, he, too was given the piece of wood.* This was how the birds learned English labels. In her scientific publications, Pepperberg always studiously avoided the word *words* for the sounds that the parrots knew. Alex, who had been working at this the longest of the three, was the most advanced and had acquired about one hundred labels. (In his classes, Alex actually preferred women teachers over men, even tall, blond ones. Alex may have thought the men were trying to dominate him, Pepperberg said.) Pepperberg was also investigating the visual systems of parrots by testing Alex's ability to perceive optical illusions—such as two parallel lines joining together in the distance—as we humans do.

Not so long ago, most scientists would have chuckled at these types of studies. Birds have small brains after all and, like fish, were thought to be missing the key neural areas that make cognition possible. Until the late twentieth century, most bird behaviors were considered innate and unchanging. For nearly three decades Pepperberg had struggled against this misperception, and the cost showed in the tiny lab she had to rent and in her relieved smile when she opened a letter and found a check for a generous donation to the Alex Foundation tucked inside. It would help her cover the expenses of the lab and parrots.

Pepperberg explained that she was between grants. Her last one had run out about eight months before I met her; she did not have a job, other than teaching one psychology class at Harvard, and sometimes had to get by on unemployment benefits. "I'm living on tofu these days," she said. She gave a wry smile and shrugged.

Throughout most of her career, she had struggled to keep her project with Alex and the other parrots going, a problem that many researchers working on long-term projects with animals face. Pepperberg, though, thought her troubles were tied to her specific study animals—parrots—and her efforts to get inside their minds by communicating with them. "It makes too many

---

* This is a very condensed version of Pepperberg's actual model/rival technique. There are many other subtle steps in the entire process, which Pepperberg describes in her 1999 book, *The Alex Studies*.

mainstream scientists uncomfortable," she said, "so getting grants has always been a challenge." Yet over the years she'd received impressive awards and grants, including several from the National Science Foundation and a Guggenheim Fellowship. Because these rarely lasted more than a year or two, Pepperberg, who'd given up a tenured university professorship to accept what turned out to be a short-lived position in the Boston area, seemed constantly to be scrounging for funds as she was now. She'd started the Alex Foundation, a nonprofit organization, in 1991 in part to help her get through the lean times.

Pepperberg had been working with Alex since 1977. She'd bought him at Noah's Ark, a Chicago-area pet store, when he was about one year old. She'd let the store manager pick out a bird for her so that scientists couldn't accuse her of choosing a particularly smart bird. Given that Alex's brain was the size of a shelled walnut, most researchers thought her study would be futile.

"Some scientists said to me, 'How can you possibly do an experiment like this? Parrots don't have brains.' People actually called me crazy for trying it. You know, 'What *are* you *smoking?*' Most scientists thought that chimpanzees were better subjects for this kind of work, although, of course, chimps can't speak," Pepperberg said.

The revolution in animal cognition studies was in its very early stages when she launched her project. Only the year before, Donald Griffin had published his book *The Question of Animal Awareness,* which opened the door to the idea that animals have minds of their own. Pepperberg boldly switched fields (at the time, she was completing a PhD in theoretical chemistry at Harvard) to join the movement, after watching two television programs. One showed scientists using sign language to communicate with apes and dolphins; the other was about how birds learn their songs. The shows "were a revelation to me," she later recalled in her book *Alex and Me.* "I had no idea what I would do nor how I would do it . . . but I did know—instantly and inescapably—that this was where my future lay." Never a quitter, Pepperberg finished her PhD in chemistry—while also attending as many animal behavior and language acquisition classes as she could, and reading every animal communication study she could lay her hands on. After being awarded her doctorate, she didn't stop to rethink her "new calling" either. Instead of looking for a chemistry-related job, she puzzled over which animal she should study.

Most researchers in the young field of interspecies communication were working with primates and cetaceans because they are closely related to humans or have large brains and, at the time, were thought most likely to have some form of humanlike communication skills. But Pepperberg, who as a child had trained parakeets to speak a few words and had kept parakeets as pets throughout her college years, thought she would have more success with a parrot.

"Parrots, especially African grays, are very good mimics," she said. "There were a few studies about the social lives of parrots and how they probably use their complex calls to keep track of the members of their flocks. So, although I hadn't raised a parrot before, I thought it made more sense to work with an animal that could vocalize"—even though parrots are far removed from the human family tree.

Pepperberg also discovered that German scientists had worked with African grays in the 1950s and 1970s. They had found that these parrots had a surprising understanding of numbers and quickly learned human speech by interacting socially with people.

Anyone who has a pet parrot knows that they readily imitate human sounds. That doesn't mean that they understand the sounds that humans make; what parrots can do is to learn to associate certain sounds with certain behaviors. If you always say, "Good-bye" when you go out the door, for instance, your parrot may come to associate that sound with your action of leaving. Your parrot may even say good-bye as you exit the door. You don't, however, expect to have a conversation with your parrot, and neither did Pepperberg. She did think that if Alex understood the association of sounds with certain objects, she might be able to ask him some questions about his perception of the world.

Over the years, some scientists have taught chimpanzees, bonobos, and gorillas to use sign language and symbols to communicate, often with impressive—if controversial—results. The bonobo Kanzi carries his symbol-communication board with him so he can "talk" to his human researchers, and he has invented combinations of symbols to express his thoughts.

Nevertheless, this is not the same thing as having an animal look up at you, open his mouth, and speak spontaneously.

Pepperberg and I walked to the back of the room, where Alex sat on top of his cage, preening his pearl gray feathers. He stopped at her approach and opened his beak.

"Want grape," Alex said.

"He hasn't had his breakfast yet," Pepperberg explained, "so he's a little put out."

Alex closed his eyelids halfway, hunched his shoulders, and looked at her. His narrowed eyelids and hunch made him look crabby.

"Don't look at me like that," Pepperberg said to him. "See, I can do it, too." She narrowed her eyes and gave him a stony look, imitating his expression. Alex responded by bending his head and pulling at the feathers on his breast.

To me, she said, "He's in a bad mood because he's molting, and sometimes when he's like that he won't work." She spoke to Alex again, "You'll get your breakfast in a moment."

"Want wheat," Alex said.

Arlene Levin-Rowe, the lab manager, handed Pepperberg a bowl of grapes, green beans, apple and banana slices, shredded wheat, and corn on the cob. Pepperberg held up the sliced fruits and vegetables for Alex, who seized them with his beak. Sometimes he held them with a claw and tore them into smaller bits. If he didn't want something, like the green beans, he said, "*Nuh*," meaning "No." It was an emphatic "Nuh"—short, and decisive. His voice had a slightly nasal and digitized quality, but it was also tinny and sweet, like the voice of a cartoon character. It made you smile.

Under Pepperberg's patient tutelage, Alex had learned how to use his vocal tract to imitate about one hundred English words, including the sounds for all of the foods she offered him, although he called an apple a "ban-erry."

"Apples may taste a little bit like bananas to him, and they look a little bit like cherries, so Alex made up that word for them," Pepperberg said.

Alex could also count to six and was learning the sounds for seven and eight.

"I'm sure he already knows both numbers," Pepperberg said. "He'll probably be able to count to ten, but he's still learning to say the words. It takes far more time to teach him certain sounds than I ever imagined."

Alex was also learning to say "brown." As a kind of learning aid for "brown," Pepperberg placed a small wooden block painted chocolate brown next to Alex.

After breakfast, Alex preened again, keeping an eye on the flock. Every so often, he used his claw to pick up the toy block and held it aloft as if showing it to everyone in the room. Then he opened his beak: "Tell me what co-lor?"

"Brown, Alex. The color is brown," Pepperberg, Levin, and the other assistant replied in a kind of singsong unison. They stretched out *brown* into almost full two syllables, emphasizing the "*br*" and "*own*."

Alex listened silently. Sometimes he tried part of the word: "*rrr . . . own*." Other times, he again held up his block and repeated his question: "What co-lor?" And the trio of humans replied together: "Brown, Alex. The color is brown."

Then Alex switched to the number seven: "Ssse . . . none."

"That's good, Alex," Pepperberg said. "Seven. The number is seven."

"Sse . . . none! Se . . . none!"

"He's practicing," she explained, when I asked what Alex was doing. "That's how he learns. He's thinking about how to say that word, how to use his vocal tract to make the correct sound."

It sounded a bit mad, the idea of a bird willingly engaging in lessons and learning. But after listening to and watching Alex, I found it difficult to argue with Pepperberg's explanation for his behaviors. She wasn't handing him treats for the repetitious work or rapping him on the claws to make him say the sounds.

"He has to hear the words over and over before he can correctly imitate them," Pepperberg said, after she and her assistants had pronounced "seven" for Alex a good dozen times in a row. "I'm not trying to see if Alex can learn a human language," she added. "That's not really the point. My plan always was to use his imitative skills to get a better understanding of avian cognition."

In other words, because Alex was able to produce a close approximation of the sounds of some English words, Pepperberg could ask him questions about a bird's basic understanding of the world. She couldn't ask him what he was thinking about, because that was beyond his vocabulary, but she could ask him about his understanding of numbers, shapes, and colors. To demonstrate, Pepperberg carried Alex on her arm to a tall wooden perch in the middle of the room. She then retrieved a green key and a small green cup from a basket on a shelf. She held up the two items to Alex's eye.

"What's same?" she asked. She looked at Alex nose-to-beak.

Without hesitation, Alex's beak opened: "Co-lor."

"What's different?" Pepperberg asked.

"Shape," Alex said. Since he lacked lips and only slightly opened his beak to reply, the words seemed to come from the air around him, as if a ventriloquist were speaking. But the words—and what can only be called the thoughts—were entirely his.

Prior to Pepperberg's study, scientists believed that birds could not learn to label objects. Assigning labels to items was something that only humans could do, linguists such as Noam Chomsky had argued in the 1960s. Scientists were also certain that birds could not understand concepts such as "same" and "different," or "bigger" and "smaller." Yet for the next twenty minutes, Alex ran through his tests, uttering the labels for a range of items (key, cup, paper) and distinguishing colors, shapes, sizes, and materials (wool versus wood versus metal) of various objects. The concept of "same/different" is considered cognitively demanding. It required Alex to pay attention to the attributes of the two objects and to understand exactly what Pepperberg was asking him to compare—their color, shape, or material. He had to make a mental judgment and then vocally give her the answer, using the correct label.

Next, she and Alex moved on to some simple arithmetic, such as counting the yellow toy blocks among a pile of mixed hues. Animals' ability to count is a much debated subject, but Alex seemed able to do this (and Pepperberg had published several papers attesting to his skill). He even understood the concept of zero, or none, as he called it—again, the only animal, other than two chimpanzees, so far known with this ability.

And, then, as if to offer final proof of the mind inside his bird brain, Alex spoke up. "Talk clearly!" he commanded, when one of the younger birds Pepperberg was teaching mispronounced the word *green*. "Talk clearly!"

"Don't be a smart aleck," Pepperberg said, shaking her head at him. "He knows all this, and he gets bored, so he interrupts the others, or he gives the wrong answer just to be obstinate. At this stage, he's like a teenage son; he's moody, and I'm never sure what he'll do."

"Wanna go tree," Alex said in a tiny voice.

Alex had lived his entire life in captivity, but he knew that beyond the lab's door there was a hallway and a tall window framing a leafy elm tree. He liked to see the tree, so Pepperberg put her hand out for him to climb aboard. She walked him down the hall into the tree's green light.

"Good boy! Good birdie," Alex said, bobbing on her hand.

"Yes, you're a good boy. You're a good birdie." And she kissed his feathered head.

DID ALEX UNDERSTAND what he was saying? From what I saw and heard, it seemed that he did. Watching him work was certainly amusing (it was hard to stop smiling at the sound of his voice) and also deeply perplexing (what's really going on here?). Pepperberg understood my bafflement—she'd encountered it numerous times. Most of us, after all, have never met an educated parrot.

"Yes, he understands what he's saying," Pepperberg said in response to my question. She had published statistical studies that proved this.

When Alex asked about the color of his toy block, Pepperberg said, he was "asking for information." And to figure out the number of yellow blocks there were on the tray, he had to stop and "actually count" the blocks. His ability to understand the concepts of same and different showed he could handle abstract thought as well. These are all higher-level cognitive skills, the kind of mental talents that prior to Alex most scientists thought only certain mammals—humans and possibly apes—could manage. He could even handle tasks beyond the capabilities of primates, such as understanding something about phonemes. Once when Pepperberg ignored his request for a nut, he finally sounded out the word for her: "Want a nut. *Nnn . . . uh . . . tuh.*"

Alex's skills, though, made perfect sense to Pepperberg. Like the great apes (and humans), parrots have long lives and are members of complex societies. And like primates, these birds must keep track of changing relationships and environments.

"They need to be able to distinguish colors to know when a fruit is ripe or unripe," Pepperberg said. "They need to categorize things—what's edible, what isn't—and to know the shapes of predators. And it helps to have a concept of numbers if you need to keep track of your flock and to recognize the various repetitions of a call. For a long-lived bird, you can't do all of this with instinct; cognition must be involved."

In other words, parrots aren't born with brains that contain nothing but preprogrammed neurons; they must learn about their world, including how to live in a society. They must consider their actions and know how to evaluate the relationships of the other parrots in their flock.

As for his ability to enunciate the phonemes of "nut," Pepperberg said that too made sense because parrots must discriminate among the calls of all their fellow parrots. A flock of parrots squawking together may sound like cacophony to our ears, but it is not just noise to the parrots.

For most of Alex's life, other scientists were unsure what to make of Pepperberg's study. Many thought she must be giving Alex hints, even if inadvertently, to speak or solve problems, and dismissed her research. They thought that Alex must be the latest version of Clever Hans, a horse that had duped the experts. In the early 1900s in Germany, Clever Hans became a vaudeville sensation by answering questions about numbers by tapping his hoof. His owner unknowingly tilted his head slightly when Hans reached the correct number. It was only a fraction-of-an-inch tilt, but Clever Hans always spotted it—while a committee of thirteen eminent scientists who investigated the horse's mathematical talents did not. Eventually, a young experimental psychologist detected the owner's unintentional cues. Hans was indeed clever, but not because he could do math.

Pepperberg had fought against her Clever Hans skeptics by making her videotapes of her sessions with Alex available, but ultimately she realized she would likely never convince them. Critics complained, too, that her study really involved only one bird, and what could possibly be concluded from a sample size of one? She had hoped that her trio of parrots would silence some of these complaints, but Griffin and Arthur are not the fast learners that Alex was (possibly because they have to share the attention of trainers, whereas Alex was an "only" parrot for fifteen years). And now Alex is gone, leaving her with but two parrots, neither of which seems to be a superstar.

This is not to say that all scientists dismissed Pepperberg's studies. Many others—from strict ornithologists to comparative psychologists—found her method of communicating with Alex, and what she'd learned about his mind, to be remarkable, particularly his ability to understand abstract concepts. They regretted that Alex had died so young, and at a time when birds' cognitive abilities were finally being recognized.

"Alex was an extraordinary parrot," said Alex Kacelnik, a behavioral ecologist at Oxford University, who was researching New Caledonian crows. "He opened the door for many of us to start thinking about birds in a new way."

Kacelnik had started his New Caledonian crow study in the late 1990s. When I visited him in 2006, he had twenty-three birds in his aviary, all but four of them caught in the forests of that South Pacific island. Although crows, like parrots, are good mimics, it wasn't their communication skills that intrigued Kacelnik but their talent as skilled toolmakers and users. In the wild, they sculpt a variety of probes and hooks from twigs and leaves to poke into rotting logs and the crowns of palm trees, where fat grubs hide. Like hunters armed with spears or arrows, the crows also carry their tools with them when out on foraging expeditions.

To show me how the crows use tools, Kacelnik and I stepped inside the aviary with a young, captive-born female named Uek (pronounced *Weck*). She was housed in a large enclosure made of wire fencing, with indoor and outdoor areas, tree limbs for climbing, and sticks, leaves, and kids' toys to investigate. Uek looked like a standard, glossy black crow. She was perched on one of the stumps but immediately flew to the ground when we stepped inside.

Kacelnik spoke to her softly, "Hello, Uek. We have a visitor."

Uek cocked her head and looked up at us, but then her attention reverted to ground level—and my running shoes. Kacelnik had told me that Uek loved to explore the seams in people's shoes. So I wasn't completely surprised when she picked up a twig about the diameter of a barbecue skewer and hopped toward me. She held the twig at an angle in her beak and deftly poked it along the seams of my left running shoe, searching for something tasty. When that proved grubless, she hopped to my right shoe and tried again. No luck there either. She was so adept with her stick—her manner reminded me of an efficient, pen-wielding secretary—that I was sorry I didn't have any resident shoe grubs.

Uek was the daughter of Kacelnik's most famous crow, Betty. Betty had recently died from an infection. But in the year before her death she had stunned Kacelnik and his colleagues by inventing a new tool during a test they'd given her. Kacelnik closed the aviary door, and we walked upstairs to his office, where he played a video of Betty making her discovery. In the film, Betty flies into a small room. She's glossy black like her daughter, with a crow's bright, inquisitive eyes, and she immediately spies the test before her on a plastic tray: a glass tube with a tiny basket lodged in its center. The basket holds a bit of meat. The scientists have placed two pieces of wire in the room. One is bent into a hook, the other is straight. They figure Betty will choose the hook to lift the basket by its handle.

But experiments don't always go according to plan. Another crow steals the hook before Betty can find it. Betty is undeterred. She looks at the meat in the basket and then spots the straight piece of wire. She picks it up with her beak and pushes one end into a crack in the tray. She then grasps the other end of the wire with her beak, and pulls it sideways—causing the end inserted into the crack to bend into a hook. Thus armed, she lifts the basket out of the tube.

"This was the first time Betty had ever been in this situation," Kacelnik said. "But she knew she could use it to make a hook and exactly where she needed to bend it to make the size she needed."

Kacelnik had given Betty other tests, each requiring a slightly different solution, such as making a hook out of a flat piece of aluminum rather than a wire. Each time, Betty invented a new tool and solved the problem. "It means she had a mental representation of what it was she wanted to make. Now that," Kacelnik said, "is a major kind of cognitive sophistication."

Scientists were astonished in the 1960s when Jane Goodall first reported her discovery that chimpanzees make tools. At the time, toolmaking was thought to be one of the key differences separating humans and animals. As amazing as her report was, it also seemed to make sense: chimpanzees and humans share a common ancestor that used tools. But our last common ancestor with birds was a reptile that lived about three hundred million years ago. How can we explain the discovery of these ingenious behaviors in creatures so far removed from our evolutionary lineage?

"This is not a trivial matter," Kacelnik said. "It means that evolution can invent similar forms of advanced intelligence more than once—that it is not something reserved only for primates or mammals." In other words, creativity and inventiveness, like other forms of intelligence, are not limited to the human line.

Kacelnik had published his discovery of Betty's inventiveness in *Science* in 2002. His was one of a growing number of studies that helped dismantle the notion that birds lack brains—a bias that stemmed from the research of the German neurobiologist Ludwig Edinger, the founder of comparative anatomy. In the late nineteenth century, Edinger dissected the brains of fish, amphibians, reptiles, birds, and mammals, noting the differences and similarities. He thought the various brains had evolved in a linear, progressive fashion, one building on the basic layers of the other, like the geological strata of a mountain. Humans, with the most layered brains, were at the peak. To Edinger, the mammalian cerebral cortex—that massive outer layer with its multiple folds of "gray matter"—was the site of all higher intelligence. Made up of a six-deep layer of cells, the cerebral cortex sits atop older, more "primitive" brain structures, such as the basal ganglia. Edinger thought that animals without a neocortex—meaning invertebrates, fish, amphibians, reptiles, and birds—could not possibly be intelligent or have any thoughts. His idea persisted throughout much of the twentieth century. As late as 1977—the year that Pepperberg acquired Alex—the comparative anatomist Alfred Romer wrote that the brains of birds were dominated by their "basal nuclei" and thus were "essentially . . . highly complex mechanism[s] with little learning capacity."

But, as with fish, anatomists had simply misinterpreted the way that the brains of birds (and amphibians and reptiles) are organized. By the 1990s, anatomists realized that all vertebrate brains consist of the same basic parts (a hindbrain, midbrain, and forebrain) and that there are structures in the brains of birds, fish, and amphibians that are homologous to the mammalian cortex. Finally, in 2004, after spending several years reevaluating the anatomy of avian brains, a team of international experts officially declared that birds *do* have the neural anatomy for thought.

No longer hampered by Edinger's bias, scientists in recent years have discovered a remarkable variety of cognitive talents in birds: Clark's nutcrackers have tremendous memories; they can hide up to thirty thousand seeds and find them six months later. Rooks, close relatives of crows, are highly inventive, making and using tools in captivity, even though they don't do this in the wild. Magpies and parrots have a sophisticated understanding of the physical world. At a very young age, they realize that when an object disappears behind a curtain it has not vanished—an ability that children also develop as toddlers. Magpies can recognize themselves in a mirror as well, an ability that suggests they are self-aware. Crows and pigeons can recognize and discriminate among human faces; pigeons can also distinguish between cubist and impressionistic styles of painting. One group of birds—bowerbirds, which live in Australia and on the island of New Guinea—even have an artistic sensibility, the only animal in which this has ever been discovered. Greater bowerbirds, for instance, use the illusion of perspective (the method an artist uses to make objects in a painting look far away) when arranging piles of progressively smaller bits of glass and stone in front of their bowers, structures they build of twigs and decorate to attract females for mating. And Kacelnik and others studying the New Caledonian crows have now shown that these crows are able to use various tools in the correct sequence and in the wild may have tool technology cultures that are distinct from one region on the island to another.

Some birds are also psychologically savvy. Western scrub jays understand that sometimes other jays are likely up to no good. The jays stash numerous nuts and seeds for the winter, just as nutcrackers do. If they can, jays will steal each other's caches, too. So a smart scrub jay that sees another jay watching him hide his nut will return later, alone, and hide the nut elsewhere.

"It means that scrub jays have something akin to 'theory of mind,'" said Nicola Clayton, a Cambridge University comparative psychologist who together with her husband and fellow researcher, Nathan Emery, discovered that jays possess this talent. "It seems that they understand what the other bird is thinking." Clayton and Emery have also shown that scrub jays not only know where they've stashed their nuts and grubs but when they did so—and will return sooner to retrieve "fresh foods," like grubs, before they spoil. So clever and innovative are jays, rooks, and crows that Emery dubbed them "feathered apes." (Clayton and Emery aren't simply besotted bird lovers. On one visit to their aviary lab, just outside Cambridge, England, Emery showed me the new experiments he and his students had devised to test rooks' knowledge of physics. Faced with a nut floating in a glass tube partially filled with water, would the rook know to pick up stones lying nearby and drop them into the tube to raise the water level—and the nut—to within reach of its beak? Yes!—the rook easily accomplished this and other demanding feats, including using tools in sequence, as do the New Caledonian crows. "They are much brighter than people give them credit for," Emery said, as we started back to Cambridge in his car. But as soon as we turned onto the main highway, a ring-necked pheasant broke from the hedgerow and dashed into the road. Emery slammed on the brakes, barely missing the pheasant. "Now there's a *stupid* bird," Emery said, shaking his head. "How many generations does it take before they learn about cars?")

Many birds surpass apes as communicators and possess vocal capabilities eerily akin to those of humans. Parrots, hummingbirds, and a variety of songbirds share with humans a talent for vocal learning: hearing a sound, copying it, then reproducing it. It's what Alex was doing when he was practicing the words *brown* and *seven*. Vocal-learning birds have specific genes and specialized parts of their brain for song learning, as humans do for speech. And like human infants, young songbirds have a babbling phase and learn their songs by listening to and imitating adult tutors. They also dream about new songs they're learning, replaying them in their minds, something scientists discovered by comparing the brain activities of zebra finches as they sang during the day and slept at night.* Songbirds that don't have an adult model to listen to

* Songbirds also generate new brain nerve cells throughout their lives—a finding that led human neurologists to take a fresh look at our own brains. Until recently, neurologists

will end up singing incorrectly, just as human infants with hearing disorders have difficulties speaking as adults.

Among mammals, vocal learning has so far been found only in whales, dolphins, elephants, seals, and bats. Other animals produce innate songs and calls; dogs and cats, for instance, cannot learn to bark or meow in a new way. These species are called auditory learners; sounds can have meaning to them, but they respond with only limited vocalizations.

In humans, vocal learning becomes more difficult after puberty, which is why adults find it challenging to learn foreign languages, although we never lose the talent completely. Some birds, such as zebra finches, learn their songs as chicks and sing these same ones for life. But a few, including mockingbirds and parrots, are lifelong vocal learners like us—and so may offer the very best models for studying this talent.

It's still a mystery as to why vocal learning evolved in such a diverse group of animals. Each group—from parrots to dolphins to humans—apparently developed it independently, just as flying evolved separately in birds and bats. But it's not an unsolvable mystery, says Erich Jarvis, a neurobiologist at Duke University, who led the 2004 reexamination of avian brains. Physiologically, he thinks, vocal learning stems from a common neural pathway for controlling motor behaviors. Behaviorally, it's surely connected to sex, he believes—to the drive to find a mating partner. "Of the species that produce learned song, all of them will do it in mating interactions, including humans."

Vocal learners often have a sense of rhythm, too, but auditory learners don't. Alex liked to bob his head to the beat of disco music from the 1980s. Snowball, a sulphur-crested cockatoo, became a YouTube sensation for his head-bobbing and Rockette dance-kicking routine to the Backstreet Boys' tune "Everybody . . . Rock Your Body." In their rhythmic displays, we readily

---

believed that adult human brains were "fixed," incapable of adding new neurons. But in 1983, S. A. Goldman and Fernando Nottebohm showed that canaries readily sprout fresh neurons as they learn new songs. Their discovery sparked a paradigm shift and led to an entirely new field of science: neurogenesis, or the study of how the adult brain generates new brain cells, something that (happily) we humans, too, are capable of throughout our lives—although, apparently, we don't do it quite as well as do birds.

recognize ourselves—and we don't think for a minute that either we or they are doing something mindless when tapping our toes to the music.

DURING MY VISIT with Alex, he noticed when his fellow parrots were slacking off during a lesson. Then he would shout at them, "On the tray! On the tray!"—because Pepperberg held up things for them to look at on a small, round tray. It was his way of saying, "Pay attention!"—something that in hindsight many scientists now wish they had done more of while Alex was alive.

As a young bird, Alex had suffered from a lung infection that may have left him in a weakened state but without any visible health problems. The evening before he died, he exchanged his usual good-byes with Pepperberg, telling her as she turned out the lights, "You be good. I love you."

"I love you, too," she replied.

"You'll be in tomorrow?"

"Yes," Pepperberg said, "I'll be in tomorrow."

That night Alex's heart gave out; a lab technician found him lying on the bottom of his cage in the morning. He wrapped Alex in a cloth and took him to the chief veterinarian at Brandeis, who placed his body in a walk-in cooler. Later, Pepperberg and her lab manager, Levin-Rowe, tucked his cloth-wrapped body into a carrier and drove him to the clinic of his regular veterinarian. Pepperberg chose not to view Alex, who over the years had become so much more to her than a colleague. He had been her friend. She wanted to remember Alex like that, as her buddy, a pal "full of life and mischief," amazing the world of science, "doing so many things he was not supposed to do." Through her tears, she whispered only "Good-bye, little friend," then turned and left the clinic.

Most African grays live into their fifties. Alex was only thirty-one—and he died only three years after the brains of birds were finally discovered. Could we have learned more from him if scientists had not been blinded by the bird-brain bias? Certainly, he offered us the potential for insights into the minds of animals in a way no other creature ever has. "Clearly, animals know more than we think and think a great deal more than we know. That essentially . . . is what Alex taught us," Pepperberg wrote after Alex died.

Still, Alex was a captive bird, raised by and with humans. And that left me wondering, what *do* parrots do in the wild that requires so much thought? Pepperberg had suggested some reasons, such as their need to distinguish among fruits and to keep track of the various relationships in their flocks. But Alex's mental abilities hinted at the possibility that parrots are capable of much more. To find out, I tracked down Karl Berg, an ornithologist at Cornell University (now at the University of California in Berkeley). He invited me to Venezuela to listen to his green-rumped parrotlets talk—not to the humans—but to one another.

4

PARROTS IN TRANSLATION

The animals want to communicate with man, but Wakan-Tanka [the Great Spirit] does not intend they shall do so directly—man must do the greater part in securing an understanding.

BRAVE BUFFALO, A TETON SIOUX

It was the rainy season—springtime—on the Venezuelan llanos when I joined Karl Berg at the Hato Masaguaral cattle ranch, where he studies parrots. The llanos are a Nevada-sized region of broad grasslands dotted with dense clumps of trees and palms. There are also swamps, lakes, and wide meandering rivers, creating the mix of habitats that make the llanos a paradise for birds and birders. Cecilia Blohm, the ranch owner, had provided Berg with a small cinderblock house for living and studying, and as we walked from the house to his field site, birds of all colors and sizes busily strutted through the green, flowery grasses and zipped among the trees, chirping, squawking, and singing sweet melodies. Falcons, hawks, and vultures soared high overhead. We were far from any city or major highway, so there were no urban sounds to mask the chorus of birds. They filled the air with their pretty, high-pitched trills and fancy warbles—songs the males sing and that scientists say are mainly about two things: attracting mates and keeping intruders (other males) out of their territories.

Isn't there anything more that birds can—or need to—say to each other?

Are all their calls and musical notes really only about sex, violence, and alarms, as ornithologists often say?

Berg was here to try to answer that question, at least for parrots. He wasn't interested in getting the parrots to talk to him. He wanted to know what they were saying to one another. He was trying to compile what Aldo Leopold once said he "fervently" wished for as he listened to a jabbering gang of thick-billed parrots in Mexico: a parrot dictionary.

To that end, we were both bristling with the gear you need to spy and eavesdrop on parrots: binoculars, spotting scopes and tripods, a shotgun microphone and recorder, earphones, cameras and zoom lenses, extra batteries, aluminum and canvas folding stools, and umbrellas to protect us from the tropical sun. The various pieces of equipment were draped over our shoulders and around our necks and stuffed into day and shoulder packs along with snacks and water. We carried the delicate scopes on their tripods in our hands.

We didn't have far to walk, Berg said, as we headed up a one-lane dirt road shaded by mango trees. He was explaining that scientists could finally begin to understand parrot calls—which are not songs—because of the latest statistics and sophisticated computer programs, but he interrupted himself to ask if I'd ever seen a vulture's egg. His question took me by surprise because until then he'd been inundating me with such densely wonkish talk about number crunching and statistical correlations that I was beginning to wonder if he liked birds or merely thought of them as data.

No, I'd never seen a vulture's egg, I replied. So Berg put aside his morning's work for a few moments and led the way through a dark grove to an old mango tree with a large cavity in its trunk. There, lying side by side at about eye level, were two blue eggs as smooth and glossy as opals. The vultures flew to a nearby branch and eyed us nervously as we admired their treasures. We didn't stay long so as not to worry the parents, but our short visit seemed to animate the birder in Berg. Beyond his aptitude for statistical wizardry ("I was supposed to be an accountant and take over my father's accounting business," he told me later), Berg was clearly—and passionately—a full-blown naturalist.

At first, as an undergraduate at the University of North Florida, Berg stuck with the family plan and majored in business and economics. Then, for fun in his senior year, he took a course in field ornithology. "I was hooked," he said, by the birds, their songs, and simply being outdoors, not in an office. Instead of continuing in economics at graduate school, he joined the Peace Corps and spent the next decade working on bird conservation projects in Ecuador. And now, at age forty-two, here he was in Venezuela, recording and analyzing parrot calls for his doctorate in the Department of Neurobiology and Behavior at Cornell University. Tall, lanky, and freckled, and with a wide-brimmed, battered straw hat shading his red hair and blue eyes, Berg looked like a grown-up Tom Sawyer. He had a wry sense of humor, too, and liked to poke fun at himself. He joked that sometimes he felt as if he'd never left the accounting world because he spent so much time entering numerical data about his birds in little spreadsheet boxes. That's what he did in his evening hours. But during the day he was out in the field as we were now, watching and listening to the birds he'd come to love the most, parrots.

Berg's years in South America had made him something of an expert on the birds of the neotropics, and as we headed from the grove toward the ranch's meadowlike pastures he called out the names of the species swooping and singing around us. I'm not a serious birder, but inspired by the avian abundance and Berg's enthusiasm I felt compelled to make a list: scarlet ibis, squirrel cuckoo, scaled dove, great egret, saffron finch, Venezuelan troupial, blue-gray tanager, dusky-capped flycatcher, and green-rumped parrotlets—the parakeet-sized parrots that Berg studies. (Because green-rumped parrotlets are a species within the Psittaciformes, or parrot family, one can also refer to them simply as parrots, as Berg often does.) He'd actually already seen dozens of the parrotlets that morning, but he waited for my sake until he spotted an entire flock just as they landed in another mango tree. They flew in with such speed that all I could see was a blur of green, as if someone had tossed a handful of emeralds at the tree. Perched, they chirped and squawked, cackled and warbled, whirred and chirred—a boisterous mix of chaotic sounds to me. But parrotlet talk to Berg.

"I'm trying to find out if parrots really do talk," he said, explaining that this was a flock of single males. "Most people say, 'Well, all those calls are just noise,' or 'they're just mimicking each other.' I think they're doing much more than that. I think they're having conversations. That might help explain some of Alex's abilities."

Berg admired Irene Pepperberg's research with Alex because "it really helped open a window on the mental abilities of parrots." Still, in his view, studying captive parrots can only tell us so much about why these birds have evolved such a talent for vocal mimicry.

"People have wondered about this for centuries," Berg said. In captivity, he added, parrots do not simply react when humans speak to them (as dogs, cats, chimpanzees, and other animals do); they also articulate responses, almost as if talking back, and sometimes even use words in the correct context, as Alex did. "Those kinds of vocalizations absolutely send a shiver up the spine of cognitive scientists," Berg said, because they suggest that parrots have some innate understanding of the purpose and function of words as sounds that convey meaning. When a pet parrot uses the words *hello, good night,* or others appropriately, it is probably not communicating about sex or violence. It is calling—and, most important, apparently meaning—"hello" and "good night."

In the wild, of course, parrots don't use the words of humans; even those that live near people have never been heard to imitate human speech. They don't copy mechanical or natural sounds, either, unlike certain species of bowerbirds and lyrebirds that are famous for their eclectic repertoires (which can include perfect renditions of sounds as diverse as the bark of a dog, the buzz of a chainsaw, or the more musical notes of other, unrelated songbirds).

So what do parrots in the wild imitate?

"Each other," Berg said. "They copy each other's signature contact calls," sounds the parrots make that serve as their names. "It took a long time to figure that out, because parrots are extremely difficult to study in the wild."

Why would parrots do this, imitate each other's names?

"Yes, that's the next question, and it's a big one," Berg said. "Imitating the calls of other parrots must help them in life—with finding food, mates, and nest sites—but how? We know they spend a lot of time, energy, and brainpower doing this, so it has to have some purpose, some reproductive benefit. But what? It's one of the things I'm trying to understand."

Berg's study of the parrotlets' calls had led him into realms he'd never imagined exploring as a field ornithologist: the evolution of childhood, child psychology and development, and the study of how humans acquire language. How, after all, did a human infant learn to speak? How did a nestling parrotlet, which emerges helpless and almost dumb from its egg, learn its calls? Are there any parallels between the two?

Like humans, male and female parrots are lifelong vocal learners. They learn their calls, as we learn to speak, by listening to and imitating others. A parrotlet chick makes its first tiny sounds just moments before hatching and emerging from its egg. At two weeks of age, the chicks begin making begging calls. At four weeks, and just before they fly from the nest or fledge, they're uttering noises that Berg regards as something akin to babbling in human infants. Only a week prior to this "babbling" stage, each chick acquires its own signature contact call: its name. "We're trying to figure out where that name comes from," Berg said. Is it innate, that is, genetically determined, and so basically preprogrammed in their brains? Or is it learned? A previous study of a different species of captive parrotlets suggested that unspecified family members assign the contact calls to each other.

Berg had an experiment under way to find out whether parrotlets in the wild do this, too, and why. From the test's results, he thought he might even discover exactly which family members assign the contact calls, the chicks' names. "One possibility," he said, "is that the parents are naming their chicks, like we name our kids."

If this proved to be the case, it would be the first time that scientists had discovered "naming" in a species other than humans. The parrotlets—and likely many parrot species in general—might then serve as a model for how and when human infants acquire language, something researchers have never had. "It would be very cool if parrots turn out to be a good model for these studies," Berg said. None of our living primate relatives, not even chimpanzees, fill this bill because they are not vocal learners.

During the current field season, Berg needed to record the contact calls of fifty specific parrotlets that were involved in his experiment—or at least as many of the birds as he could find. That's what we would be doing for most of the next few days. He hoped I didn't get bored, because searching for particular birds was "pretty much like looking for a needle in a haystack." Secretly, I was amazed that anyone would try to do this: hunt for fifty individual birds, each the size of a parakeet and the color of a leaf. Even if the birds were banded, in this wide-open terrain finding them seemed like a very long shot. I hoped to see just one. At that moment, Berg came to a standstill. He focused his binoculars on a parrotlet zipping past us.

"Hear that peep?" Berg asked. "That's one of our [study] birds, and that's his signature contact call—his name. It's what the other parrotlets imitate."

One of our study birds!? I didn't stop to ask Berg how he knew this, or how one had turned up so suddenly, because I wanted to see the bird for myself and hear his call. I wanted to hear a parrotlet say his name.

"*Peep . . . peep . . . peep . . . peep . . . peep.*"

"You mean *that* 'peep'?" I asked, as I studied the parrotlet through my binoculars. I was puzzled and a little disappointed, although I tried not to show it. The parrotlet's signature contact call—his name—was as short and soft as the peep of a newly hatched chicken.

"Yes, that peep; that's what I'm studying," Berg said, adding, "I know what you're thinking. The first time I heard it, I thought, 'Are you kidding? I'm not studying *that*. I mean, how much more boring a sound could a bird make? But I also thought it wouldn't take me very long to figure out, and that I'd probably get the cover of *Science* by showing how the birds use it and what it means. But," he said, drawing a deep breath, "birds call and sing at a speed that we're not equipped to hear, so we don't hear all the variation that's really in that peep. Figuring it out hasn't been as easy as I thought."

Although Berg had already published some of his research in leading scientific journals and also received a coveted ornithological award, he knew that some people might consider him foolish for spending years studying what seemed to our ears such a tiny, insignificant sound. He stuck with the project because of the rare opportunity the birds presented: if he could tease apart the details in the parrotlets' peeps—separate their names from other possible information—he would be well on his way to unraveling their chatter overall. What, after all, is the purpose of a name? Think for a moment about how and why you call your friends by name. "Hey, Jack," you say, hoping to get his attention by calling (that is, by imitating) his name. The method works. Jack turns to look at you. He responds by calling out yours, "Hey, Jill." Now you're ready to start a conversation.

CAN ANIMALS CONVERSE with their own kind? Are their communications or vocalizations in any way akin to human language? Darwin thought they must be, since we can understand the cries and gestures of monkeys and the barks and expressions of dogs. Although he did not spell out how human language evolved, he thought it must have come about through one of the great forces of evolution, natural selection. "Man has an instinctive tendency to speak," he observed, "as we see in the babble of our young children." In other words, the human language ability is grounded in biology and has a biological history, even if we like to think that we speak with the tongues of angels.

Until recently, however, most scientists shied away from the question of the origin of human speech. Although Noam Chomsky, the most influential linguist of the twentieth century, had argued as long ago as 1975 that the human brain is endowed with a "language faculty," he also thought it had appeared *de novo:* that is, it was not a result of natural selection even if it was hardwired and was therefore unrelated to the communicative abilities of other animals. He argued that because it was impossible to do more than speculate about language's evolutionary roots, the subject wasn't worth studying. Chomsky's opinion held until the early 1990s. By then, the wave of evolutionary thinking that swept through the cognitive sciences was also engulfing linguistics. New discoveries in genetics, the neurosciences, and brain imaging techniques made it possible to look at the biology of language in ways never before possible; and new findings in animal communication research suggested surprising and unexpected links between animal vocalizations and language. In 2002, even Chomsky stepped back from his previous ideas and, in a seminal paper coauthored with evolutionary biologist W. Tecumseh Fitch and psychologist Marc Hauser, urged researchers to consider the evolutionary—meaning biological—roots of language. Scientists should try to separate those aspects of language that are uniquely human from those that are shared with other animals, the trio wrote in *Science.*

Linguists, however, are far from accepting the notion that anything like language has been found in another species. Human language is still generally considered unique in the animal kingdom, a special adaptation along the lines of an elephant's trunk or a bat's sonar. Linguists say that animal calls lack the key elements of true language: the ability to use abstract symbols, such as words, in an infinite variety of ways to communicate about the past, present, and future. In contrast, animal vocalizations seem to be solely about the present—*this* moment, *now.* They are also largely exclamatory: I want a mate! I see food! I see an enemy! This part of the neighborhood is mine! Keep out! And they are repetitive. As lovely as the robin's springtime lilting song may be to our ears, he's likely saying only: This is mine! Keep out! This is mine! Keep out!

Most linguists also insist that animal calls lack grammar and syntax, the principles and rules that determine how words can be combined into the phrases, clauses, and sentences of language. Recently, however, some scientists who've been listening more closely to the calls of other creatures argue that certain species may have something roughly analogous to the rules of language.

In the forests of the Ivory Coast, primatologists studying male Campbell's monkeys have successfully translated the monkeys' cries announcing that they've spotted predators. Other researchers have decoded similar alarm calls in other primate species, prairie dogs, meerkats, and chickens. But Campbell's monkeys also seem to have something like syntax, or "proto-syntax" as the primatologists call it—that is, they add extra sounds to their basic calls to change the meaning. We do this when we change the word *neighbor* to *neighborhood.* Campbell's monkeys have three alarm calls: *Hok* for eagles, *krak* for leopards, and *boom* for nonpredatory disturbances, such as a branch falling from a tree. By combining these sounds, the monkeys can form new messages. A Campbell's monkey who wants another monkey to join him calls out, "*Boom boom!*" That means "I'm here, come to me!" report the French and British team who've recorded and studied these vocalizations. "*Krak krak!*" can be translated as "Watch out, a leopard!" But when the monkeys combine the two calls—"*Boom boom krak-oo krak-oo krak-oo*"—they mean something entirely different: "Watch out, a falling tree!" Simply by adding that "oo" sound they effectively double their repertoire, so that *Krak-oo* functions as a general warning, alerting others to almost any disturbance, while *Hok-oo* tells a monkey's fellows that there's a commotion in the canopy, such as a perched eagle or an opponent from a rival group. The *oo* serves as something akin to a suffix in human language, the scientists say. The monkeys also call back and forth, varying and altering the sequences—apparently having something like conversations, or at least exchanges of information, just as Berg suspects the parrotlets are doing.

Although Berg has not yet translated any of the parrotlets' specific vocalizations (other than their contact calls), his initial work shows that they also vary and combine their calls, and in even more complicated ways than the monkeys. The parrotlets also learn to articulate their own names and those of other parrotlets, something monkeys cannot do because they are not vocal learners. Thus, although many species of monkeys have contact calls, and although monkeys know one another's calls, they don't use these as the parrotlets do—as names.*

All of which suggests that the vocalizations of the parrotlets, and all parrots, may offer the closest parallel yet to human language. This is not to diminish the primatologists' achievement. Translating the calls of other animals is a hugely complex task. The scientists deciphering the Campbell's monkeys' calls have spent close to a dozen years learning the monkeys' natural history, getting them used to humans, observing their behaviors in multiple situations, recording their calls, and analyzing these in a lab—all in order to tease apart six sounds. For the scientists involved in such work, it's rather like being a member of a long-distance marathon or relay team—except that the pace is like a slow walk, and there is no obvious finish line.

BERG PICKED UP THE PARROTLET-CALL marathon baton in 2003. He inherited the project from Steve Beissinger, an ecologist at the University of

* Monkeys learn to distinguish each other's contact calls by listening. But they are only auditory learners. They are not vocal learners and cannot imitate another monkey's contact call.

California, Berkeley, who noticed a pair of parrotlets nesting in a hollowed-out fence post at Hato Masaguaral in 1985. "Parrots usually nest high in the treetops, which makes them extremely difficult to study, but these were nesting just a few feet off the ground," Beissinger recalled in a telephone conversation. "I immediately wondered, 'Would they nest in artificial boxes?'" Two years later, after experimenting with designs, he constructed a faux fence post from a three-foot-long piece of white PVC pipe. He lined it with wire mesh, cut a cavity about six inches from the top, and fitted removable lids on the top and bottom. He then filled the bottom of the tube with wood shavings, hung the contraption from a post at the ranch, and waited. About a month later, a pair of parrotlets moved in. The next year, Beissinger added forty more PVC boxes, and many of these were soon occupied.

With the ranch owner Cecilia Blohm's blessing, Beissinger made 106 such nesting boxes, spacing these about thirty feet apart among the regular fence posts. It was as if a developer had turned the fences at Hato Masaguaral into a green-rumped parrotlet condominium heaven. The parrotlets flocked to the nesting boxes. For the scientists, it was like having an enormous, outdoor wild parrot lab. Although Beissinger didn't realize it in 1987, when he set out the first nesting box, he had launched the world's longest-running study of wild parrots.* "There is nothing like it anywhere else for studying parrots in the wild," Jack Bradbury, a parrot expert and Berg's advisor at Cornell, told me. "It's a phenomenal system" because the birds can be monitored and handled experimentally throughout their lives. For their studies of the parrotlets' calls, the scientists can also place video cameras and recorders inside the boxes to track the birds' physical and vocal development from the moment the eggs are laid until the chicks fledge. When I visited in 2009, the project was entering its twenty-second year, making it the parrot equivalent of Jane Goodall's long-term study of chimpanzees.

---

* One of Irene Pepperberg's students, Diana May, attempted to study African gray parrots in the Central African Republic and Cameroon, but gave up after three frustrating field seasons, a coup, and encounters with machete-wielding parrot poachers. May was able to collect some recordings of their vocal behavior and other data, which have not been published, although her video of the poachers has been useful to parrot rescue organizations.

Beissinger's original research interest was in the parrotlets' ecology and their basic behaviors, simply because so little was known about wild parrots and because parrots are one of the most threatened groups of birds in the world. "One-third of New World parrot species are at risk of extinction due to poaching and habitat loss," he said. "Any data we collected, I thought, might help." (As it did; some of the team's findings have influenced the drafting of regulations that govern how wild birds can be imported to the United States.) Beissinger and his team studied the parrotlets' demography, social system, and "hatching asynchrony," which means that the siblings in a parrotlet's nest range in age from a few days to two weeks old. Over the next sixteen years, the scientists gathered data on every aspect of the parrotlets' lives by banding and tracking thousands of the birds, documenting three thousand nesting attempts, and following the fates of sixteen thousand eggs.

By the time he handed over the project to Berg for his vocalization studies, Beissinger had amassed genealogical and genetic data on more than 8,500 parrotlets, along with all the intimate details of their life histories. He recorded the data in two ledger-style books that are stored at the ranch. (The scientists also transcribe the data into computer databases.) "The books are like our Bibles for the parrotlets," Berg said. He set them on the dining table one evening in the scientists' ranch-based lab, a cluttered all-purpose room that also served as their kitchen, and gear closet. "You know, who begat whom."

Turning the pages, Berg explained how, in the main ledger, researchers keep track of those biblical begats so that they can determine each bird's genealogy and pedigree and understand its complex familial relationships. When designing an experiment, scientists try to control for as many variables as possible, something that is difficult to do with a population of wild animals, where often little is known about their ancestry. The parrotlets' genealogies are as detailed as you'd expect to find in a laboratory setting, where it's standard practice to record lab animals' pedigrees and life history events. The genealogies make Berg's experiments with the wild parrotlets possible.

Out in the field, Berg showed me how he and his colleagues add to these records every day, while monitoring the nesting boxes. We placed most of our gear on the ground close to a nesting-box fence line. Then we walked its length, opening a dozen boxes and peeking inside to check for signs of nesting and eggs. The first box Berg inspected was empty, but in the second an ivory-hued egg about the size of a cherry lay nestled in the wood shavings. Berg pulled a felt-tip pen from his pocket, picked up the egg with the care of someone handling a rare jewel, and gently marked it with a number. Later, he would enter the number in the ledger. He hoped we would see chicks, but the parrotlets had delayed their nesting season, apparently because the rains were late. If we had found nestlings, he would have weighed and measured them and outfitted their tiny legs with a set of plastic colored bands and an aluminum one bearing an identification number—more codes that would be entered in the ledger. The scientists use the ID numbers and colored bands to track the birds throughout their lives. It isn't easy, though, to decipher through a spotting scope or binoculars the colors of a parrotlet's rings, which are the size of miniature lifesavers, and Berg's talent at this task never failed to impress me.

After checking the nesting boxes, we went back to our gear and set up our stools and scopes so that we sat facing a nesting box, marked #104. The week before, Berg had discovered one of his naming-study birds, Male-7358, nesting here with his first-ever wife. Berg wanted to collect both parrotlets' signature calls, while keeping an eye and ear out for other parrotlets. He had a clipboard with a stack of worksheets for gathering data about the birds' behaviors, which he would enter that evening in another book, the *Resighting Log*. The scientists monitored the birds this way every morning and afternoon. They sat with their spotting scopes, as Berg and I were doing, near a nesting box, identifying any nearby parrotlets by the color bands on the birds' legs, and marking their observations on the worksheets. They penciled in a code for each bird's behavior. Thus, if a parrotlet was starting to nest in a box, Berg entered the letters NE for "start of nesting"; if a parrotlet was at a box with a bird of the opposite sex, he wrote BO for "at box with opposite sex"; if it was alone at a box, he wrote BA for "at box but alone." Some codes suggested a darker side to the parrotlets: AR indicated the parrotlet had behaved aggressively toward another member of the flock; AE, that a parrotlet had been the aggressee, the one being attacked. DS was a code for "suspected to be dead"; DD confirmed that a bird was "definitely dead"; and BW denoted a male who was courting a widow—"at box with widow."

So is the *Resighting Log* like a parrotlet version of *Desperate Housewives*? I asked.

"Every box has a story," Berg said. "My wife tells these best; she's Ecuadorean, and says the birds have a lot of Latin drama in their lives." Soraya Delgado, Berg's wife and a fellow ornithologist, was in the States, however, so Berg related the *Log*'s parrotlet tales. Perhaps they weren't as spicy in his telling, but they were plenty hair-raising, with sagas involving wife beating and infidelity, trickery and divorce, nesting-box theft, murder, and even infanticide. He agreed that it wasn't easy reconciling all this social mayhem with the sweet-faced, chubby-cheeked birds we were watching through our scopes, but, as he said, "It certainly gives them something to talk about. Sometimes, I think that's what most of their calls are about: gossip." Some scientists, such as Robin Dunbar, have proposed that the original purpose of human language was gossip; he imagined protohumans picking lice from each other's hair while passing on news about the troubled family down the way. Certainly, the parrotlets lived eventful, gossip-worthy lives. There wasn't only family drama; there were predators for the parrotlets to worry about, too. Black boa constrictors slithered into nests, killing mothers and kids; falcons picked off foraging dads.

Berg didn't always bother with strict scientific terminology when telling the birds' tales but described the parrotlets as he'd come to see them—as husbands and wives, moms and dads, widows and widowers, and gangs of bachelors, who he said sometimes behaved like "thugs."

Because the parrotlets' signature calls are extremely plastic—that is, they vary the sound slightly depending on whom they're contacting (mate, chick, or friend)—Berg thinks they might be adding some other information to a parrotlet's name. Perhaps a pair attach a sound of affection when they call each other's names. Or, if they think their neighbor is a jerk, they may say his name along with a note of disapproval. Or, if a neighbor has been killed by a snake, they may add sounds denoting fear or death and snake to her name. It could be a useful way to pass on news from the neighborhood: "Sue, kids, dead, snake." Steven Pinker and others have suggested that human language emerged for exactly these reasons—as a means of passing on knowledge to others.

Many of the dramas begin with the birds competing for a nest. The parrotlets know what constitutes a good nesting box, and they know which ones are the most successful—that is, where pairs have successfully raised their chicks. They avoid the boxes that are closest to thick vegetation, where predators can hide, and they fight for those "in the best neighborhoods," as Berg put it. "There are never enough nesting cavities. Without a box or natural hollow, they can't raise a family. Once a pair gets one, they have to defend it; a lone parrotlet can never do this on its own." (The parrotlets' need for a box may help explain why they tolerate the researchers' handling them, their eggs, and chicks. Even such disturbances are not enough to get them to give up those boxes.)

Some couples want a particular box and will even kill to get it. The gangs of young bachelor males (those thugs) hang around the boxes, always on the lookout for unhappy couples or weak-looking husbands. They harass the nesting couples, dive-bombing and pecking the husband to drive him away. There is always the chance, after all, that if he dies or leaves, the widow will choose one of them for her next mate. What happens if she already has a family? I asked.

"Well, any eggs and kids are at risk," Berg said. The new dad may kill them, which is also certain to happen if another couple comes along and drives the widow from her home. That is a terrible thing to see—the cute little chicks with their punk-rock, spiky pin-feathers, bloodied and torn to bits; their bills ripped from their faces; the eggs smashed. Those scenes break Berg's heart. Occasionally, a new dad surprises—and puzzles—him by adopting a widow's chicks. ("We can explain infanticide; the new male doesn't want to raise some other guy's kids. That's not going to help his reproductive success. I don't have an answer yet for the adoptions.") He's seen lots of happily affectionate pairs, too, that kiss, feed, and preen each other. They sing and dance together as part of their courting rituals and, after mating, snuggle, while twining their bills. Happily married pairs may have sex right out in the open, sometimes on top of their boxes—perhaps as a way of "sending a message to the neighborhood that we've got this box, and we're tight," Berg said.

Parrotlet sex isn't just the usual five-second cloaca-to-cloaca "kissing" that most birds do, either. Berg thinks the males may have an intromittent organ, probably about the diameter of a thread, that engorges with fluids for mating, much as a mammalian penis does. He has videos of parrotlets having sex and showed them to me later. He warned me beforehand that parrotlet sex looked surprisingly humanlike, and it did. Pushing up against her, the male used a claw to clutch one of the female's wings and held her tight to his chest. Then, with their faces close together, he mounted her from behind and actively thrust for several minutes. "My goodness, Karl," said Berg's Venezuelan research assistant, Malu Gonzalez, as she and I watched together in disbelief, "I didn't know you had *that* on your computer."

Berg didn't collect the tales of parrotlet sex and drama simply for the sake of entertaining visitors while watching for particular birds to show up at their nesting boxes. The scenes he had witnessed and the stories he culled from the *Resighting Log* held clues about what was important to the birds—about why they make particular calls in particular situations and why they need names.

"They really do use their contact calls the way we use names," Berg said. "It's a very effective social tool. Little in our society would function without the use of our own names and the ability to imitate the names of others." He pictured the female down in the bottom of her dark nesting box; she's waiting for her mate to bring her dinner, but she can't see out. How does she recognize him? "She listens for his signature call. She doesn't come up unless she hears *him;* she won't surface to the sound of another male's call. We know; we've done the experiment. We've played the recorded contact calls of strange males—and the females don't come up. They're doing something similar to what we do when we recognize a friend's voice over the phone; we picture them mentally. It's an ability we take for granted, but it really requires some sophisticated mental processing."

Berg turned in his chair, and pointed to a calling parrotlet that was sitting on a wire between two fence posts. Right behind the bird was an acacia tree, where a small flock of parrotlets was perched. "Do you hear that male? He's calling to his buddies, his friends, in that tree."

I focused my spotting scope on the male, and his emerald green face popped into view. In most parrot species, males and females look alike (which is another reason they are difficult to study in the wild), but female parrotlets have a bright splash of yellow above their beaks, while males have rich turquoise markings on the leading edge of their wings, and turquoise and lapis markings beneath their wings—in their wing-pits. When trying to attract a female, they flash their colors by holding their wings aloft, then snapping them shut—rather like a guy opening his jacket to give the ladies a quick glimpse of his pecs. The females choose their mates (as is the case in most species), so they may be paying attention to and evaluating differences in color and shape (who has the brightest blue feathers?), just as Alex did.

"*Peep, peep, peep, peep, peep,*" the male called. His wings were folded demurely against his body, so he wasn't trying to impress any parrotlet females. How, I wondered, did Berg know he was calling to his friends? Often, as we walked around the ranch, Berg would say things like "There's a male and female pair singing a duet in that acacia tree" or "There are two guys in that bush, they're fighting about something" or "I think that guy just called his buddy's name." I found it difficult simply to hear the peeps, they were so soft and fleeting.

Berg nodded. "It's like hearing a language; it takes time to get your ear tuned to their calls. I had trouble at first; now I hear them all the time here." Sometimes after he returned from the field to the States, his brain was so full of parrotlet calls, he even heard them there—or thought he did. "Part of the problem is also that they just call so fast; they can make twenty calls in the time it takes you to sneeze. And we really can't make sense of their peeps, or any of their calls [they also make chirps, warbles, and threatening growls] just by listening—even if you slow them down."

Later, after we were back in the States, Berg sent me a recording from his lab of a parrotlet's *peep* that he had slowed down so that I could hear the sound as a parrotlet might. Transformed in this manner, it no longer sounded remotely like *peep*. It had become a weird concatenation of almost guttural consonants, with a few higher-pitched vowel-like notes thrown in here and there. In my notebook, I wrote: *eh-eehhhhh-gehhhllll—grrrrr-whoeeeee*. Berg chuckled at my rendition. "Well, that's why we don't study them by trying to write down what they sound like; it really doesn't work."

Scientists get a better idea of what a parrot's call or bird's song sounds like via a spectrogram—a visual image, like a musical score, that depicts the frequency, timing, and amplitude of the call. As we waited at box 104 for Male-7358, Berg pulled out a sheet of paper with a printed spectrogram of a parrotlet's contact call. It resembled a Chinese brushstroke: a broad line that swept upward to a peak, then dipped slightly downward—*peep* rendered as a graphlike picture.

Berg had converted thousands of peeps from hundreds of parrotlets into spectrograms. He then ran these "sound-images" through specialized computer programs that searched for and quantified subtle similarities and differences in the calls. This was how, in part, the scientists had discovered that each parrotlet has a unique contact call and that they can imitate each other's calls.

"They're not just calling out each other's names, though," Berg said. His study of several mated pairs showed that males and females make fifteen basic calls to each other, which they link together in various sequences. Sometimes a mated pair exchanges calls rather like songbirds singing duets, with one rapidly vocalizing a series of peeps and the other imitating that series. "But it's not a song," Berg emphasized. "It's a dialogue or a conversation—in the sense that what I say influences what you say next. That's what the parrotlets are doing."

While answering my questions and waiting for birds to show up, Berg usually sat with his arms folded over his lap or rested his chin in one hand. He was, perhaps not surprisingly, an attentive listener, but I also realized that his eyes and ears were constantly scanning for signs of his birds. When he heard one or caught sight of one, his focus immediately shifted, as it did now. He sat up, reached for his microphone, and trained his spotting scope on box 104.

"I think that's our bird, the dad," Berg said, as a parrotlet flitted along the edge of the pasture, heading toward the box. It was Male-7358 and he was peeping. Berg held his shotgun microphone aloft to capture the male's call, and his mate's. She poked her head out of the box and peeped; he peeped in response. The two exchanged a couple of more calls, and then she ducked back down inside, where he joined her. He would feed her there.

"I still don't know what those exchanged calls mean," Berg said. "But it's clearly not just 'Hi, honey, I'm home.' And 'Hi, how're you doing?' He could be saying his name and 'I'm home and I've got plenty to feed you. But I want sex first.' And she might be saying, 'Well, I need to eat first. Then we'll have sex.' There's some negotiating going on."*

Male-7358 stayed only a few minutes inside the box before flying off on another foraging trip to gather more seeds. He needed to work hard, Berg said, to keep his mate well fed so that she could continue laying eggs. About ten minutes later, he was back, peeping to her while in the air and then as he landed on the box. This time she joined him outside. They sat side by side on their nesting box, calling back and forth. We expected to see Male-7358 feed her. Instead, he made a quick, darting move and bit her on the back of her neck.

"Ouch!" Berg said, as the two flew off hurriedly together. "Well, there's some marital strife in that nest. He could be a bad husband. She might end up divorcing him."

Berg thought he would eventually be able to translate these exchanges by playing back short sequences of his recordings to the parrotlets and recording their responses, as well as noting how the birds behaved in response to the calls. He had used this type of playback experiment to show that the parrotlets have signature calls that others recognize. Thus a parrotlet listening to a playback of his partner's call often responds by mimicking the call—an exchange that might be translated as "I'm Sylvia." . . . "Yes, I hear you, Sylvia." But sometimes in these experiments, parrotlets that hear their mate's call don't respond at all. Why not? "We clearly don't know everything that these calls are conveying," Berg said. "We could be playing a sequence that says, 'Shut up,' or 'Be quiet, there's danger.' We just don't know."

BERG STARTED HIS CURRENT EXPERIMENT in 2007. Guided by their detailed genealogy records, he and his assistants swapped complete clutches among twelve nests, so that the chicks were raised by unrelated foster parents, who

* Jack Bradbury, an emeritus professor and Berg's mentor at Cornell, suggests that parrots, which live in dynamic flocks, may be using their vocal mimicry to help negotiate flock separations and mergers. For instance, orange-fronted conures exchange contact calls in late afternoon when recruiting others to their sleeping roosts.

had different calls from those of the biological parents. They used eight other nests as controls; these nestlings lived with their biological parents. After the chicks hatched, Berg made weekly video and audio recordings inside the nests, as well as outside during the parents' arrivals and departures.

By the end of the nesting season, fifty chicks (including Male-7358) from the original seventy-six eggs had flown the nest; three nests had failed. Berg had gathered the contact calls of twenty-five cross-fostered chicks and twenty-six control chicks, along with those of the parents. In the Cornell Lab of Ornithology, Berg began comparing the spectrograms of five thousand such calls and analyzing them statistically. If the chicks' calls are innate, they should resemble those of the biological parents rather than those of the foster parents. He was still in the early stages of this complicated study when we met in Venezuela, but he was already excited by what he could see.

"We just grabbed some of the calls at random, and wham! there it was—the cross-fostered chicks' calls match those of their foster parents," Berg said. "They don't sound like their biological parents. And that means the chicks are, at the very least, learning their calls."

So the peep of Male-7358 (that is, his name) was one he'd learned from his foster parents?

"Yep," Berg said, "that's what it's looking like." (Two years later, in June 2011, Berg published the results of this experiment in the *Proceedings of the Royal Society B*. As he suspected, the chicks' calls most closely resembled those of their foster parents; the calls were not strictly inherited but learned.)

Did this then mean that the foster parents had named Male-7358, that is, they had given him his signature call?

"Maybe," Berg said. He rubbed his legs. He didn't want to make extravagant claims; he wanted to wait, like an accountant, for the final analysis and statistics to back him up. But as we talked, Berg allowed that this was a plausible explanation for how the chicks acquire their calls.

"It makes the most sense, especially when you consider how important the calls are to everything they do, right from the beginning." Like humans, he explained, parrotlets have extended parental care, with parents feeding their fledglings for another three weeks after they leave the nest. The young birds roost in large, communal groups with as many as three hundred other recently fledged chicks, "which makes it just that much harder for the parents to find their own kids," he said. "They'd never manage without the contact calls." Beyond being born dumb and helpless, there are other similarities between human infants and the parrotlet chicks. Both have extended dependent childhoods, too, which allow for the development of large brains. Also like humans, the parrotlets hit a key developmental milestone when they begin to learn their names; in humans, that stage is a sign that the infant is able to begin connecting words with meanings. Perhaps the parrotlets reach an analogous stage when they begin learning their signature contact calls.

If so, then, Berg would have uncovered an "incredible evolutionary convergence" between parrots and humans, something that not even Aldo Leopold, who longed for a parrot dictionary, could have imagined.

FOR AS LONG as the cross-fostered chicks survive, Berg hopes to track them and record their calls, building on this one simple experiment. Collecting the calls of Male-7358, for instance, and other now-mature parrotlets in his study, would begin to show something about how the birds' calls change over time. What do they add to their repertoires, and why?

Unquestionably, the calls help reduce the uncertainties in the parrotlets' lives, Berg said. He used Male-7358 as an example. "His big job right now is feeding his wife, and then the chicks when they hatch." But it is a risky job, because parrotlets are seed eaters, and they have to hunt for seeds on the ground, thereby exposing themselves to danger. They can be snatched by raptors or snakes. By foraging with friends, a male has a better chance of avoiding one of these predators. Males, like 7358, build up their friendships over the two years it takes them to reach sexual maturity (the females are ready for mating at six months of age), while hanging out in the large, all-male flocks. A lot of all that "noise," Berg said, "is simply contact calls: 'Hey Bob! Hey, Jack! Hey, Joe! You're my bud.'

"It's all about individual recognition, and communicating information to your friends, and not to others," he continued. "Well, who are your friends?

The ones whose names you know. The ones you like to warble and forage with."

And the females? They may be attracted initially by a fellow's bright markings, but ultimately they may need to pick a mate who has a large repertoire—that is, the guy with the long list of friends. "Females are smart," Berg said. "If they choose a male with a lot of buddies, it may be that they're less likely to end up as widows."

Apparently, Male-7358 had an extensive enough repertoire because a female had chosen him as her mate in his first year of eligibility.

Half an hour after Male-7358 and his partner departed together, they zoomed back to their box. They landed on its flat top and sat for a moment side by side. This time there was no biting, but no canoodling either. Berg was still not convinced that this marriage would last. Then she dropped over the edge and silently ducked inside the nesting cavity. He lingered for a few minutes and finally flew away, making a few peeping calls in time with each wing dip.

Berg held his microphone aloft and recorded the calls until Male-7358 was out of sight.

Who were those calls directed to? His wife or his pals?

"I'd say his pals. He needs to go foraging. He needs to find his friends."

Maybe he had called some of his friends' names and added a trill for foraging and another for their favorite meadow—whatever the snippet of conversation was about, Berg had it stored in his recorder, another few peeps waiting for translation.

ON MY LAST DAY at the ranch, Berg and Gonzalez set up several mist nets to capture a male and female pair who were not yet banded but had moved into a nesting box. It didn't take long before the male flew into the net. Gonzalez quickly freed him. Two more parrotlets, whose bands needed replacing, were also caught. She put each bird in a cloth sack to keep them quiet and took them to Berg, who had set up a small field lab in the shade of a tree.

Gently, Berg weighed and measured the birds and fastened the bright bands on their legs. He called out the data to Gonzalez and helped her take blood samples from the parrotlets for genetic testing. He held the male for a moment longer, carefully stretching out a wing to show me his fancy blue markings. The little parrotlet opened his beak—not to bite Berg but to emit a call. Another male parrotlet had landed in the tree above our heads and was sitting as close to us as he could and still be safe. He answered the parrotlet Berg was holding, giving a high-pitched peep.

Berg smiled. "Is he your buddy?"

"Did he respond with the other male's call?" I asked.

Berg raised his eyebrows. "Maybe," he said. "It's what we like to think—what we're trying to show."

What would a parrotlet trapped in an alien creature's hand say to his brave friend? I wondered. Did they have expressions of fear and worry and danger?

Perhaps, Berg said, but he was years from tackling that kind of detail. "Sometimes I think I'll be old and gray before I get even the basics of their conversations worked out."

Holding the newly banded male, Berg stood up and let him go. The male flew off, making quick calls: *peep, peep, peep, peep* . . .

"I guess we gave him something to talk about," Berg said.

AT ITS HEART, Berg's study is about the importance of social skills in acquiring and fostering intelligence—a notion that was first put forward by the primatologist Alison Jolly and the social psychologist Nicholas Humphrey in two separate papers in 1966 and 1976. Both argued that the complexities of social living—of understanding what your neighbors were doing and possibly plotting—were the key evolutionary pressures in developing complex cognitive abilities. Their papers led to the field of social cognition. The more demanding the society, the more pressure there would be for intelligence to evolve.

How do animals go about building social skills? The male parrotlets seem to rely heavily on their calls to acquire their necessary friends. But just as Berg doesn't fully understand what causes a female to choose one male over all the others, he has yet to figure out how the males make friends, how they select the ones whose names they add to their repertoires and go foraging with. "That's a mystery," he admits.

In other species, such as baboons and chimpanzees, friendships are common, too, but are better understood because they've been studied far longer, and scientists have devised various ways to measure degrees of friendship and the strength of alliances. Grooming, picking out lice from another's fur, and assisting your pal in fights are two of the best ways in the primate world to make a friend. And then there are the white Norwegian rats found in every university's psychology department. They're adept at making friends, too, but they do it by playing and laughing together. When I first read about giggling rats, I was doubtful, although it also seemed to make sense. After all, what could be more social than sharing a laugh?

5

THE LAUGHTER OF RATS

Although some still regard laughter as a uniquely human trait, honed in the Pleistocene, the joke's on them.

JAAK PANKSEPP

By his own reckoning, Jaak Panksepp has tickled more rats than anyone else in the world. An emeritus professor of psychobiology at Bowling Green State University and currently a highly regarded neuroscientist at Washington State University in Pullman, Panksepp pioneered the field of what he calls affective neuroscience—the zone where neurons, emotions, and cognition meet. How, after all, does the brain generate emotional feelings, and how do these feelings affect an animal's behavior?

Panksepp and his laughing rats have helped overturn the old, Cartesian idea that emotion and reason are separate entities. Today, emotion and cognition are acknowledged to be inextricably intertwined—at least in humans. Some researchers are still reluctant to assign anything more than just a few emotional behaviors to other animals. But for Panksepp, emotions and their associated feelings are "evolutionary skills" that are constructed in the brains of all animals to help them face the challenges of life; they are animals' "genetic birthrights." His studies, which are widely cited by scientists investigating animal welfare issues, have influenced the humane approach Temple Grandin, the celebrated autistic author and scientist, advocates for managing animals and have also led to new treatments for autistic and attention-deficit children.

Although Panksepp emigrated from Estonia to the United States when he was six, he still has a trace of his accent, due to his parents' rule that their children speak their native tongue even in their new Delaware home. His ski-slope nose, round face, and trim ring of white hair and beard give him the kindly look of a Christmas elf. On this morning, he was wearing his special set of cat scent–free clothing—charcoal black trousers and a long-sleeved blue-and-white striped shirt—because he didn't want to frighten his rats. Although his rats and their laboratory-bred ancestors had never encountered a cat, they knew that feline odor signals danger. A cat's scent would make them freeze in place. They would not play; they would not laugh.

"They know something is amiss as soon as they get a whiff of that odor," said Panksepp, who owns a cat, as we made our way to his lab in Washington State University's Veterinary College. "I don't know how many other researchers using rats realize this, but rats still have an innate wariness of cats"—even though most lab rats haven't had to worry about becoming some feline's meal in nearly 150 years. But rats have been fleeing from cats for millennia, and it will take more than a 150-year safety-zone interlude to erase their deeply rooted cat alarm.

Panksepp had spotted the rats' fear at the outset of his play experiments, which he launched in the early 1980s. Some days, for no apparent reason, the rats behaved more cautiously, cowering close to their playpen's walls. Guessing that the rats' fear might be because he and some of his other cat-owning colleagues had the felines' odor on their clothing, Panksepp put a tuft of cat's fur in the rats' play area one day. The rats immediately stopped their roughhousing. Now he keeps his special cat-free clothes stored in a plastic bag and in a room at his home that his cat is forbidden to enter.

Since I also own a cat, I had taken my own precautions to not upset Panksepp's rats, putting my jeans and blouse through two wash-and-dry cycles, packing them in plastic bags, and not petting my kitty Nini good-bye.

A sign on the double doors to Panksepp's lab commanded: QUIET, EXPERIMENT IN PROGRESS. From a cupboard next to the door, Panksepp retrieved two sets of blue nylon booties. We fitted these over our street shoes so that we wouldn't track dirt and bacteria into the lab.

We were ready to meet the rats.

We entered a tidy, windowless room lined with metal shelves, each holding white plastic bins topped with barred grates. Even before I saw the rats, I heard them: that telltale, rustling sound of rodent claws as the animals bustled over wood shavings. All were Long-Evans rats, which scientists had bred in 1915 from the wild *Rattus norvegicus*, or Norwegian rat (which is a misnomer; they were mistakenly thought to have entered Europe from Norway but in fact came from Southeast Asia). Long-Evans rats have more genetic variation than most lab rat strains, and Panksepp regards them as a model for the basic, "original" mammal, akin to the first rodent that appeared about seventy-five million years ago.

Sheri Six, Panksepp's lab manager, was just starting the day's experiment. In her early thirties, Six has short, brown hair and blue eyes and a no-nonsense manner; the experiment might be about play and laughter, but it was part of a serious study, and Six went about setting it up with brisk efficiency. Ten of the twenty rats in her care would be allowed to play for thirty minutes with another rat, while the other ten stayed alone in their bins. The solitary rats were always alone. They had never played with another rat—and never would. With nothing to do, the solitary rats were all asleep, curled into tight furry balls in their corner nests, even though Panksepp said that this was the age in a rat's life—their fourth and fifth weeks—when rats most enjoy playing.

The rats that got to play knew it was their playtime. They weren't in bed. Like kids waiting for their playdates, these rats were up and ready, their black eyes bright with anticipation. They poked their pink, whiskered noses through the gates, sniffing and twitching, and didn't struggle when Six gently lifted them from their bins.

"We're trying to understand the function of play," Panksepp said, raising one of those issues that scientists have battled over since Darwin's day. Because playing doesn't appear to be connected directly to an animal's survival—to its need for food and to reproduce—some researchers argue, like Calvinists, that playful behavior serves no purpose. Others see it as a necessity for learning survival skills, such as hunting, escaping from predators, and mating. Panksepp takes that idea one step further.

"We argue that play—especially rough-and-tumble play—helps construct the social brain," he said, his "we" referring to his lab's colleagues. "It lets the rats explore the limits of fear, anger, lust, and care, and other subtle feelings, such as exploratory seeking. It makes them confident."

By electrically stimulating the brains of rats and guinea pigs, Panksepp has mapped and defined seven fundamental emotional systems found in the mammalian subcortex. He calls these FEAR, RAGE, LUST, CARE, PANIC/GRIEF, PLAY, and SEEKING—and uses all caps to spell them to emphasize that they are scientific terms. Each one represents a specific system in the brain and not simply the sensation, say, of fear or lust. The seven networks serve similar functions in all mammals, from rats to humans.

Those twitching rat noses that poked up at us? That was a sign of the rats' confidence, a trait that, Panksepp believes, they had acquired through playing. "Sniffing is part of their SEEKING system," he said, referring to a circuit in the subcortex of the mammalian brain, a region that also governs such vital functions as hunger, thirst, breathing, and sleep. SEEKING is the impulse to explore one's environment and world, and it showed in the rats' eagerness for something new to happen.

Six picked up Rat-71, a male with ermine-white fur, slipped him into male Rat-72's bin, and closed the grate. Rat-71 and Rat-72 had been playing together twice a day for more than two weeks. They sniffed each other for a moment, perhaps to verify that "yes, you're my play pal," then got down to business. Rat-71 put his paws on Rat-72's back ("That's a play solicitation," Six said), and the two launched into a vigorous bout of tumbling and wrestling. They chased each other around the bin, dashing one way, then making quick pivots to run in the opposite direction—just like a pair of puppies. They took turns, like young dogs, too, with one falling deliberately on his back, and the other pouncing on him, pinning him to the ground and nipping at his neck; then they reversed positions. They leapfrogged, and jumped on and off each other. They bowled each other over, and tussled and rolled in such a carefree, happy way that I found myself laughing.

Although I've watched many different animals play—from my pet dogs and cats to lion cubs and baby elephants (I've even seen carrion crows in Austria make snowballs with their beaks and slide on their backs down snowbanks)—I'd never seen a rat play. And, until this moment, I'd never thought of them as a particularly playful species. They have short lives, often no more than two years, and such short-lived species are generally more concerned with life's absolute essentials: find food; find a mate; reproduce. Yet there was no mistaking their exuberant frolicking for anything other than play.

"Little kids, too, have no trouble seeing that this is play," Panksepp said. "But some professors, when they've watched, have asked me, 'How do you get your rats to *fight* like that?' It isn't fighting; they aren't being aggressive. They are *playing*."

Some of the rats' moves, such as the nipping, might look aggressive, but Panksepp said the resemblance was only superficial. When rats are playing, they rarely square off and box each other, and they don't raise their fur to make themselves appear larger as they would in battle. Moreover, they never draw blood. Panksepp notes that "if one animal stops playing, often the other makes small, darting nips." They do the same thing when he tickles them. "If I stop, they start nipping at my hand to provoke more fun."

The rats seemed surprisingly quiet as they wrestled, jumped, and ran merrily around their bins. Occasionally, one made a short, loud squeak. Was that rat laughter?

"No; it's a play-solicitation—a request to keep playing. You can't hear the laughter; it's out of range of our hearing," Panksepp said. "A high-pitched chirp can be a happy sound. But if it's insistent, or the rat makes an audible squeal, then it's a kind of protest: 'You bit me too hard.' If the roughhousing gets too real—if it becomes like fighting—the one that's being hurt will stop playing. That's the point of play: this isn't 'for real.' It's for fun. And there are rules that keep it fun."

Some rat play rules: Pinning must be brief and must be reciprocated. Bites must be quick and not cause actual harm. In all the rat play pairs, one rat ends up winning a majority (about 70 percent) of the pinning matches—but the winner keeps the game fun by handicapping himself or herself, sometimes letting the other rat hold him or her down. "If he doesn't do that—if he acts like a bully—the other rat will stop playing," Panksepp said.

Rat-72 was the winner in this pair, and he wasn't a bully: the two rats played with scarcely a break for the full thirty minutes. The four other pairs of rats had also played with abandon. The youngest rats, numbers 81 and 83, grappled together, belly to belly, looking at times like a single furry ball as they rolled around their bin. They would jump apart for a few seconds, only to grab each other again for another round of tussling and nipping.

Partway through the session, Panksepp said that although we could not hear the sounds, he knew from experience that the playing rats were chirping. "They're making the sounds of laughter," Panksepp said. He didn't like to say they were "laughing" (although sometimes he slipped and did so), because laughing is something only humans do. "I'm sure some of them were even chirping when we first entered the room and even before they began to play. They make chirps in anticipation of play." Panksepp knew this because he and his students had made recordings of the rats as they waited for their playdates.

A room full of laughing rats! Their joyful chirps were ricocheting all around us, but we couldn't hear a bit of it. If there was a moment that encapsulated all that we don't know or miss about animals, for me, this surely was it. It was like being in a foreign country when all the locals break into big guffaws at someone's joke or quip, and you—not speaking the language—can only look on, a passive spectator.

We needed a rat-laugh translator, and for that we had a bat detector, a handheld device about the size of a GPS unit, connected to a computer. Like rat laughter, bats' echolocation calls are inaudible to our human ears because they are ultrasonic—beyond our hearing range. Bat detectors convert bats' high-frequency signals (120 kilohertz to 160 kilohertz) and rats' laughing chirps (50 kilohertz) into the lower frequencies we hear (2 kilohertz to 4 kilohertz).

Panksepp led me next door into a lab with a dozen more rats in bins. They were part of what he called his "high-line" rats, rats he and his team had bred with especially happy dispositions. He also had developed "low-line" rats, which rarely if ever made laughing play-chirps and weren't very playful; in fact, they were depressed. The high-line rats were just the opposite. They were exuberant, uninhibited play-chirpers, Panksepp said, and by comparing them with their low-line brethren, he and his colleagues are tracking down the genes for joy.

Sheri Six lifted a month-old white female high-line rat from her nesting bin and placed her in one that was empty.

"She's never been tickled," Six said, stepping back.

"Good," Panksepp said. "If she likes it, you'll see how quickly she responds."

He didn't talk soothingly to the rat, or stroke her, but firmly grabbed her behind the neck, mimicking a playful nip, and then ran his fingers up and down her rib cage, tickling her. She squirmed briefly, but stopped when he turned her over and tickled her belly. (Like humans, rats have "tickle-skin.") That was when she began to laugh, calls that we heard through the bat detector as quick, high-pitched chirps, and saw on the computer monitor in a sonogram rendition as a vertical series of wavy lines. Compared to a sonogram of various kinds of human laughs, a rat's chirps may be closest to a giggle.

"There, she's laughing already," Panksepp said, tickling her some more.

"*Chup, chup, chup,*" I wrote in my notebook, trying to approximate the bat-detector's translation of her rat laughter.

When Panksepp stopped tickling, she jumped up and bunny-hopped around the bin, while making more of her laughing play-chirps.

"Ah! Bunny-hopping already," Panksepp said. "That's a clear sign of joy. It's a move you see in rats, dogs, and other animals when they're playing and happy." When he tickled her, he was simulating play. He called tickling "faux play."

The rat stood up and peered over the side of the bin, sniffing eagerly.

"She's looking for my hand, for the tickler," Panksepp said.

He reached into the bin again, and flipped the rat on her back and tickled her stomach for another minute. When he stopped tickling this time, he kept his hand in front of her nose, quickly rubbing his thumb and forefinger together. He then pulled his hand away from her, but she followed, pursuing his fingers around the bin.

"She's hoping my hand will tickle her some more; she's seeking it," Panksepp said. "It's a sign of social bonding—she wants to be with my hand," the hand that had tickled her and made her laugh.

Panksepp gave the rat another quick bout of tickling and then asked if I would like to tickle her.

"Don't pet her or stroke her," he warned, since in previous experiments he had shown that rats much prefer tickling as opposed to petting hands. Rats will even choose a tickling hand over another hand that has only petted them and will run a complex maze to get tickled, but not to be petted.

I imitated Panksepp's moves, first squeezing the rat gently on her neck, then running my fingers along her sides.

"She seems tense," I said. She had also stopped her laughing chirps, and I wondered if she would bite.

"Yes. She knows this is a new hand," Panksepp said. "I think she'll relax." As for biting, tickled rats have nipped Panksepp hundreds of times, but never enough to draw blood; he regards these as playful nips, the way a rat nips a fellow rat—a rat's request to keep playing.

I tickled her some more along her sides, and I could feel her relax. Then it was easy to flip her on her back and tickle her tummy. Her fur was soft as a rabbit's and warm. Soon she was laughing. And when I stopped, she jumped up, sniffing for more. So I tickled her again, no longer worried that she might bite—not just nip—me. She bunny-hopped and jumped around her bin, just as Rats -71 and -72 had, seeking more fun and making several loud chirps that we could hear without the bat detector.

"That's a solicitation call for more play," Panksepp said.

I happily gave her another quick tickle, unsure at this point who was actually having the most fun.

WHY ARE TICKLING AND LAUGHING such fun for rats? And how are these and play connected to cognition?

"It's all about social joy," Panksepp said, as we walked from the lab to his office. "You and I have just been listening to the sounds of social joy in rats. Now, joy and grief and other emotions aren't just special capacities of the human brain. Other animals have these emotions, too, and they aren't invisible or impenetrable, contrary to what some researchers say." Emotions can be explored, tested, and understood empirically in other species, he explained, via artificial stimulation (with electrical probes or chemicals), as he and his students do regularly in their studies.

Throughout most of the twentieth century, animals were regarded as not having emotional feelings. They had behaviors that were emotional, such as fear and anger, but scientists denied that these corresponded to any feeling, that is, any internal, mental experience. And even if animals had feelings, these were thought to be beyond study because they were invisible—that is, not tangible, not evidenced in a way that could be empirically measured. Panksepp has long argued the opposite. "The emotional mind is the *most* visible part of an animal's brain," he said. "You can see it directly by watching the animal's behavior, and you can hear it in their vocalizations."*

In the 1940s, Walter Hess, a Swiss physiologist, showed that it was possible to make a cat angry by electrically stimulating the hypothalamus area of its brain. Hess and most other neurologists argued that the cat wasn't actually angry because nothing had happened—a dog hadn't chased or cornered it—to make it upset. Panksepp has never accepted that interpretation. "I've always believed the cat *was* angry. It behaved in an angry manner; it arched its back, it spat and hissed. It told us it was angry."

Most other researchers denied the cat its feelings, Panksepp thinks, because of that "old knee-jerk reaction, that fear of anthropomorphism," of attributing human thoughts and feelings to animals. "We've been accused of that, too, because we use the word *laughter.* But, this is not simply some 'belief' that we have; it is based on solid evidence from our neural investigations into the brains of rats."

The desire to play is so deeply embedded in the brains of rats (and all mammals) that when Panksepp surgically removed the upper brain (the neocortex) of three-day-old rats, they still played—and made the play chirps of rat laughter. "That showed that play is a primitive process," he said, adding that "the great lesson of twentieth-century molecular biology is the abundance of evolutionary continuities across species, just as Darwin taught. Laughter and play didn't appear out of nowhere. They have evolutionary roots."

All the great apes laugh, as do dogs, which make quick, panting, breathy sounds when frolicking with their owners or other dogs. Probably many other

---

* The neuroscientist Antonio Damasio reached the same conclusion: "I do not see emotions and feelings as the intangible and vaporous qualities that many presume them to be. Their subject matter is concrete, and they can be related to specific systems in body and brain."

species laugh as well, Panksepp thinks, but scientists have simply not noticed the sound or misinterpreted it. It took him five years of studying rat play chirps before he realized they could be the rat equivalent of primal human laughter. "I had an intuition one day," Panksepp recalls. "It dawned on me that those chirps were similar to the first laughs human infants make, when they're about three or four months old. And at that age, it's the child's primitive part of the brain that's responding," the basal region that is strikingly similar in all mammals and that Panksepp says is the location of our basic emotions.

Panksepp is emphatic on this point, arguing that his neural studies as well as those of his colleagues show that the prime, fundamental emotions of humans and all mammals do not emerge from the cerebral cortex, as was commonly believed in the twentieth century and as some leading neuroscientists still claim, but come from deep, ancient brain structures, including the hypothalamus and amygdala. It is why, he notes, that "drugs used to treat emotional and psychiatric disorders in humans were first developed and found effective in animals—rats and mice. This kind of research would obviously have no value if animals were incapable of experiencing these emotional states, or if we did not share them."

The rats that love to be tickled the most and that laugh the most are those that have just been weaned and then kept in isolation for at least one day. As soon as someone tickles a rat that's been made "hungry" for play, it erupts in a cacophony of ultrasonic chirps, more than double the amount other rats make, as if it had been longing to laugh its entire life. Nothing else—not the most delectable treat—elicits such an elated vocal response.

"At that age, their brains deeply desire play and laughter," Panksepp said, "because they're evolutionarily so important for optimal social growth."

Far from being purposeless, playing helps assemble the brain, Panksepp's team and others have shown. It acts like a spark, triggering the release of proteins that cause neurons to sprout and grow not only in lower emotional-memory regions of the brain such as the hippocampus and amygdala but also in the prefrontal cortex, the brain's decision-making area.* Rats that have

* Just by watching animals play, Darwin reached the same conclusion as today's neuroscientists: "We see [joy] . . . in the bounding and barking of a dog when going out to

played with each other and with toys grow more neural connections than those that are kept alone.

And at least some of those neurons are tied to the development of proper social behaviors. Panksepp's playful rats learn the best way to approach other rats, how to read another rat's intentions, and how to make friends. Rats that laugh the most are sought out by other rats; their laughter is infectious. When mature, female rats also seem to be more attracted to the most playful males. Panksepp has run some pilot tests, letting a female choose between a male with plenty of play experience and one who never played. "The play-experienced male gets the gal," he said, because this male knows how to position himself between the female and the other male without causing her alarm. "The other guy loses out completely." Other experiments in the works suggest that rats that have played may be better able to handle stress and fear than those that haven't, and may be protected against depression.

"Basically, the nonplay rats are like attention deficit disorder kids," Panksepp said. "In social situations, they're more liable to get into a serious fight. They don't know how to inhibit their reactions properly."

And that applies directly to cognition, too. "You must have control of your emotions, which playing teaches you how to do, in order to think."

DO RATS THINK? Every rat exterminator likely would say yes, since rats have managed to outwit them for centuries. Rats have a capacity for living anywhere, from ships to city sewers and subways to posh apartments. It's not just a matter of luck that rats are so successful and hardy—although over their nearly 150-year tenure in laboratories they've been regarded largely as unthinking, unfeeling machines. (How else to explain experiments such as one the psychologist C. P. Richter designed in the 1950s to see if rats would develop a feeling of helplessness if tossed into a vat of warm water without any

---

walk with his master; and in the frisking of a horse when turned out into an open field. Joy quickens the circulation, and this stimulates the brain, which again reacts on the whole body."

means of escape? And, yes, the test produced a sense of helplessness in the rats. When they realized that their efforts to escape were futile, they simply stopped swimming and just floated, having given up trying to find a way out.) The attitude that lab rats are merely living machines that lack mental experiences is beginning to change, but not fast enough to ensure that rats and mice are treated humanely. Under the Animal Welfare Act of 2004, both of these animals, as well as birds and fish, all of which are subjected to millions of experiments each year, are excluded from the definition of *animal*.

As "model" mammals, lab rats and mice are used as stand-ins for us, physically and psychologically. Increasingly, researchers are finding that rats, at least, are more like us than most of us would like to believe. Scientists at the University of Georgia recently discovered that rats are self-aware and capable of something like introspection—complex cognitive abilities long attributed only to humans, dolphins, rhesus monkeys, and apes. Rats are smart enough to know what they know—and what they don't know—and make their decisions accordingly, a mental ability called metacognition. "Rats have the ability to reflect on their internal mental states," says Jonathon Crystal, one of the University of Georgia neuroscientists (he's now at the University of Indiana) who discovered metacognition in rats by rewarding them for *not* pressing a lever in a test if they did not know the answer. Crystal has also shown that, like scrub jays and other nut- and seed-storing birds, rats remember where and when they've found food in the past and can use those memories to plan for the future.

Rats also dream as we do, reliving the day's challenges in cinematic narratives. Researchers monitoring the rats' brains during REM (rapid eye movement) sleep found the same pattern of neurons firing that they had seen when the rats were actually wending their way through a circular maze. The patterns were so similar that the scientists could pinpoint exactly where in the maze the rat would be if it were awake.

In other labs, rats have been discovered to be expressive individuals with personalities. Some are good-natured and cheerful, others glum and pessimistic. They make grimacing facial expressions when they're in pain and sigh in relief when they know they're not going to receive an electric shock to their feet (which happens in many experiments). They can be altruistic and offer help to a strange, unrelated rat. They like sex, and, according to James Pfaus, a rat-sex researcher at Concordia University in Montreal, they "know what good sex is and what bad sex is," and they look forward to the former. Since both sex and play are based on trust and require cooperation, it may be that the most play-experienced rats are also the best at sex.

Panksepp likes to say that his joyful rats may not have a "sense of humor, but they sure do have a sense of fun"—a trait far removed from any machine.

"That was a tall tale, a twentieth-century myth," Panksepp said about the animal-as-machine model. While there are scientists, particularly in his field, he added, who still believe that "emotional feelings are special capabilities found only in the human brain, it's just not so."

BEFORE PANKSEPP DISCOVERED LAUGHTER in rats, he studied grief (or PANIC as he calls it—namely separation distress calls) in dogs, guinea pigs, and chicks. (He's recently begun to study these calls in degus, small Chilean rodents.) He was the first to investigate the brain circuits of a crying baby mammal, the desperate wails young infants make when separated from their mothers, similar to the sounds of a baby that has wandered off and cannot find its way home. The calls of these animals are not ultrasonic but loud and urgent. By electrically stimulating the midbrain and various higher brain regions, he could provoke these distressed calls in baby guinea pigs and chickens, just as Walter Hess had provoked anger in cats.

The circuitry doesn't involve the neocortex, the mammalian brain's executive area. Like expressions of play and anger, separation calls are triggered in the subcortex, the more primitive region of the brain. Panksepp says the cries are evidence of the brain's basic "PANIC system"—one of those core emotions that all mammals share. Rats, hamsters, kittens, puppies, human babies—every young mammal will make such calls, so dependent are they on their mothers for survival. A lost infant mammal will sink into despair and ultimately die if not reunited with its mother. Baby chickens and other young birds also give separation distress calls, suggesting that the urge has even deeper evolutionary roots.

"A young, helpless animal is born needing to be cared for. It doesn't need to learn this or to learn how to cry if it loses its way, as the behaviorists used to argue," he said. "It is an inborn cry of psychological pain."

Baby guinea pigs, puppies, humans—we all perceive emotional "hurt" as pain because, as Panksepp also discovered in the 1970s, the brain's panic area sits close to the region for regulating physical pain. Intriguingly, when he stimulated the guinea pig's brain areas that were close to those for physical pain, the animal made separation distress calls, not pain sounds. He also found that he could turn off the pain of separation by giving the guinea pig an opiate drug. "It was exactly like treating physical pain." Intrigued by the connection between the two types of pain, Panksepp suggested that the PANIC system evolved from the area for physical pain, an idea that didn't receive much support until 2003 when it was validated by studies in another lab. Researchers at the University of California in Los Angeles used MRI scans to watch what happened inside the brains of people who were suddenly excluded from a virtual ball game. Although it wasn't a real game, the people felt they had been rejected—and the hurt of that rejection showed in what the scientists termed the "human sadness system," an area related to the separation distress circuitry that Panksepp had identified and mapped in guinea pigs' brains in the 1980s.

"We all know the pain of rejection," said Panksepp. "We say it *hurts* if we're socially rejected, or if we lose a loved one. Why do we feel this as pain? Because there is a common neurological basis for social and physical pain."

Were the guinea pig's cries simply calls that would bring the mother running? Or did the guinea pig feel pain and sorrow—grief—that it could not find its mother?

"I've always been the fool who's said animals experience the emotion," Panksepp said—not because of some amorphous "belief" but because of tests he'd performed to determine if animals "like or dislike" having parts of their brains electrically stimulated. For instance, he'd discovered that rats have much the same anger (or RAGE system) areas in their brains as do cats. When he stimulated this part of a rat's brain with an electrical probe, the rats became highly agitated and enraged. He then gave them the choice of pressing a lever that would turn the stimulus on or off. The rats turned it off.

"It wasn't a sham reaction to them; they experienced feelings of rage, which are bad and unpleasant." The same holds true for the baby guinea pig's cries of sorrow, he said. "It may not be identical to what we feel, but at its core, the feeling is the same. The distress cry means the animal is alone and lonely, a terrible thing for a baby mammal."

A mother guinea pig who hears her lost infant comes running to its separation cries, because the calls are distressing to her. Panksepp suspects there is a parallel track of pain in her brain that her youngster's calls triggers. It is why mother cats yowl for their lost kittens, and cows bellow for their calves.

Some tragedies are too cruel to be explained. As we talked, Panksepp mentioned that he had difficulty completing his major book *Affective Neuroscience*, which describes his theories about the neurology and evolutionary history of animals' emotions. He'd been unable to work for some time, he said. Had he been ill? I asked. No, he replied. In 1991, a decade after finishing his studies on the baby guinea pig's distress calls, his only daughter, Tiina, died. She was just sixteen and was killed along with three friends when a drunk driver slammed into their car. He told me this toward the end of my visit, his voice laden with sorrow, his body and face crumpling in pain. He had raised her for much of her life as a single father and knew the utter anguish of losing a child.

Panksepp has also known the sting of social rejection. While some of his studies have been lauded, many of his fellow neuroscientists still dispute his notion that the rats and guinea pigs feel similar emotions to ours. To them, an animal's separation cry is simply a reflex response, conveying no more emotion than a sneeze. They argue that Panksepp and his supporters are projecting their own emotions onto rats and guinea pigs.

"Of course," he said. "You could show them the identical circuits—even the molecules—of emotion in humans and rats, and they still would deny the similarity." But eventually, he predicts, they'll give way as the mounting body of scientific evidence for emotions in animals becomes undeniable.

Panksepp's work in the controlled conditions of a lab, studying small, easily handled mammals with sophisticated imaging tools, is one matter. But how do researchers working with wild elephants or dolphins prove that their study animals have thoughts and emotions? Without the option of MRI scans or electronic probes to map their emotional circuitry, how does one delve into the farther reaches of the animal mind?

One of the longest-running studies of African elephants in the wild was just entering its thirty-eighth year in Kenya's Amboseli National Park. There had been a drought the previous year, and many elephants, young and old, had died. Did the elephants know this, and were they grieving for their lost relatives?

What do elephants feel when tragedy strikes?

6

ELEPHANT MEMORIES

Who can know what goes on in the hearts and minds of elephants but the elephants themselves?

JOYCE POOLE

"I like to think of myself as an animal-mind detective," Karen McComb said to me early one morning in March 2010 as we drove through Kenya's Amboseli National Park in search of elephants. "My dad was a detective in Ireland, and he cracked some very tough cases. I like to think I have some of his skills. I don't just jump to conclusions at the outset. I go where the clues lead."

McComb's self-description seemed accurate. During an earlier game drive, she'd displayed a detective's eye for detail, noting subtle elephant behaviors that eluded me: the twitch of an ear, the positioning of a trunk, the hesitating step. Her eyes narrowed as she weighed what to do or say next; and she became visibly uncomfortable if pressed to make a statement about something she couldn't support with facts. At the University of Sussex, on England's southeast coast, McComb is codirector of the Mammal Vocal Communication and Cognition Research Group. More specifically, she's a behavioral ecologist and a leading expert in parsing the vocalizations of mammals to decipher their thoughts and emotions. Their reactions to their calls are her clues.

"Since my childhood, I've just always wondered why animals do the things they do," McComb said. In her early fifties, McComb has a wavy mass of blond hair, an athletic build, and the intense focus and high spirits of someone who loves her job. Over the years, her research has taken her from Cambridge, England, to Scotland, New Zealand, Minnesota, and Tanzania. Along the way, she's picked up traces of the other countries' accents, softening and rounding her original Northern Irish twang. At Cambridge University, where she studied for her doctorate in the 1980s, McComb discovered a talent for imagining what it might be like to be another animal and inventing an experiment to test her idea. She didn't focus on a single species, as some scientists such as Jane Goodall and ant researcher Nigel Franks have, but chose instead to pry into the minds of mammals in general. In addition to elephants, she's studied cats and dogs, horses, red deer, monkeys, and lions—most often by playing recordings of their calls and then watching, videotaping, and tabulating the animals' responses. She's revealed her findings in a number of landmark papers: on the complexity of male mammals' mating calls, using red deer as her example (prior to her study the calls were thought to be solely about male-male competition, but the females are listening and judging, too); the ability of horses to recognize other horses' whinnies, which means they must have mental images of the horses they know; and the hidden demands in a domestic cat's purr-cry for food (a study inspired by the behavior of her own cat). The combined purr-cry is such an effective call, McComb says, that cat owners will leap to their feet and feed the cat first—even before the dog, or before pouring themselves a cup of coffee. In animal cognition circles, she is widely regarded as a kind of mammal whisperer.

In 1990, Cynthia Moss, an ethologist and the director of the Amboseli Elephant Research Project, invited McComb to join her team. Moss's project was then in its eighteenth year, the longest-running study of elephants in the wild. During that time—and in the two decades since—Moss and her collaborators have amassed extensive observations and genealogical data on some seventeen hundred individual elephants from sixty-one families. They know which elephants are related and whom each elephant is related to. They know which elephant families spend time together, who their leaders are, how elephants react when they meet strangers or encounter a predator, and how families behave when they split apart and come back together. They have numerous eyewitness accounts of elephants helping one another and of elephants seeming to grieve for their dead.

What the team lacked, and what Moss hoped McComb would help them gather, was proof that the elephants are cognitively and emotionally engaged—that is, they are thinking and feeling—when they behave or react in a particular way.

"Elephants are a bit of a challenge," McComb allowed. "They're big, of course, but that's not such a problem. I think what makes them most difficult is that many of their social behaviors and their close family ties remind us of ourselves. So the easiest thing is to make them like humans, to assume they think like we do, even though they have a very different ancestry and body architecture from ours—just look at their trunks, and think of how many ways they use that organ. We have nothing like it." McComb wished she could be an elephant "just for one day to find out what it is really like. I probably feel that way about any animal I'm studying."

McComb had picked me up early that morning in the team's Land Rover, instructing me to wait outside my safari lodge to save time. As we motored across the park, her eager postdoctoral assistant, Graeme Shannon, a young Scot, was at the wheel, and seated beside me in the back was Norah Njiraini. A member of Kenya's Kikuyu tribe, Njiraini had joined the project in 1985, when she was in her midtwenties; she's now one of Moss's top field scientists. The day was an especially fine one for elephant viewing, Njiraini told me. The season known as the long rains had started, and everything—roads, trees, sky, even the snowy peak of 19,340-foot Mount Kilimanjaro—looked fresh. A few weeks before, the park's grasses had been stunted and brown; now they were a brilliant emerald, and already tall enough to sway in the morning's breeze. "The elephants will be getting fat," Njiraini predicted.

After a forty-five-minute drive, Njiraini spotted a particular elephant family she wanted to show McComb. Shannon drove a short distance off road to position the Land Rover so we could get a close look at the herd, and soon we were surrounded by elephants—about twenty of them, mostly adult females, youngsters of both sexes, and two one-year-old calves. Towering above us, the elephants moved with the slow, stately grace of sailing vessels crossing a green sea. Some stopped now and then to crack off leafy branches from an acacia tree, or to uproot fat bunches of green grass with their trunks. Sometimes they lifted their trunks to assess the car and its occupants. They were close enough that we could see the mud caked on their foreheads and in their long eyelashes, and smell their earthy, barnyard scent of fresh hay and manure. You might think that such a big group of large animals would cause a commotion just by walking. But except for the occasional snapping of a tree branch as they fed on the acacias, they passed by almost silently, their huge, padded feet meeting the ground as inaudibly as cats' paws.

The elephants were accustomed to the Land Rover, and to McComb and her fellow scientists. But although I'd made previous visits to Amboseli, I wasn't part of the team, and the elephants knew it, Njiraini said.

"See that one raise her trunk?" Njiraini said, pointing to a young female, who shook her head and sidestepped away as she took in our odors. "She doesn't recognize your scent. That's why she is shaking her head. Elephants don't like a strange smell. She's saying, 'Eeww, Mommy, who's that?'" Njiraini wrinkled her nose and shook her own head, imitating the elephant. Because the matriarch of this family didn't make a fuss, the youngster settled down and went back to feeding, apparently accepting that there was just going to be an unpleasant odor emanating from that usually friendly vehicle on this otherwise fine Amboseli morning.

Njiraini wasn't just spinning a tale about the matriarch's decision to ignore my smell. Moss and McComb had first collaborated on precisely this issue—the role of the matriarch. And, after seven years of experiments, they had amassed sufficient data to prove that in every elephant family the matriarchs—the eldest females, usually forty-five to sixty-five years old—are the leaders. "They're the storehouses of knowledge—about the landscape, predators, and other elephants—just as Cynthia had always argued," McComb said. "Watching these elephants, it may look like they are only wandering, but their matriarch is leading them; she's making decisions moment by moment about what to do and where to go, about what's best for their families." It doesn't matter if a family is large or small, or if it has several older females, the matriarch orchestrates most of the family's activities and actions.

Now the scientists wanted to find out how the matriarchs acquire their knowledge and how it informs their decisions. The experiment that McComb planned for this morning was part of the team's new investigation.

The elephants we'd joined didn't take long to move past us. Following their matriarch, they began walking toward an expanse of muddy flats. McComb studied them with her binoculars. She was looking for just the right herd—a family without any newborns that might get upset by the experiment—and just the right location, ideally an open plain without many bushes and trees. "I think this is going to work," she said to Shannon. "Let's drive around there on the left and set it up."

Shannon hit the ignition, and we began lurching across the savanna toward the elephants.

McComb turned back to me. "There's something else we hope to explore this season, or to reexplore. And that's the question of how elephants respond to the death of their own kind."

Twenty of the eldest matriarchs—the wisest leaders of Amboseli's herds—had died the previous year in a severe drought, the worst in decades. Did any members of their families remember them? And if they did, were they mourning?

Can scientists answer such questions? I asked.

McComb didn't answer right away. "Perhaps," she said at last. "We're going to try."

AS LONG AGO AS ARISTOTLE, people have remarked on the wit and intellect of elephants. It wasn't until recently, though, that anyone could cite much more than anecdotal evidence of elephants' superior intelligence. One obvious reason: How do you evaluate the cognition of a five-ton creature with a mind of its own? Can you even conduct tests in a controlled laboratory setting? And what tools do you need?

Joshua Plotnik, a graduate student from Emory University, faced just such questions when, in 2006, he wanted to see if three Asian elephants at the Bronx Zoo—Happy, Maxine, and Patty—were self-aware: that is, if they could recognize themselves in a mirror. First, Plotnik and his collaborators had to build a safe, jumbo-sized mirror that was up to the task. They glued two large, mirrored acrylic sheets to a plywood backing and framed the resulting eight-by-eight-foot panel in steel. Elephants in previous self-recognition experiments had failed, but those tests had employed small mirrors that were mounted beyond the elephants' reach. This time, Plotnik's test subjects passed. None of them acted as if they were looking at another elephant when they studied their reflections. They stood in front of the mirror and used their trunks to investigate their mouths, and played with it, as a child might, by moving in and out of its range, apparently watching themselves appear and disappear. Happy also used her trunk to repeatedly investigate a white mark that Plotnik painted on her head. Plotnik's conclusion: elephants, like great apes, dolphins, and several other species, are aware of themselves as individuals.

They're also one of the few animals capable of solving a problem by using insight—which is that *aha!* moment when your internal lightbulb switches on and you figure out the solution to a puzzle. It was once thought that elephants lack insight, because in one test they apparently couldn't figure out how to extend the reach of their trunks by grasping a stick and then using the stick to knock down a piece of fruit dangling overhead. But in 2010, Kandula, a young male Asian elephant at the Smithsonian National Zoological Park in Washington, D.C., was offered not a stick but a kind of step stool. Kandula rolled the sturdy cube to the middle of his enclosure, stepped up on it, and snagged a suspended treat with his trunk—thereby upending a long-standing misconception about the elephant mind, while also revealing the limitations of our own at times. "The primary purpose of their trunks is to smell," said Preston Foerder, a graduate student at the City University of New York, who devised the successful elephant-insight test. So when an elephant is asked to hold a stick with its trunk to reach food, it can grasp the stick—but it can no longer locate the food. "It would be like having an eye in the palm of your hand, and then being asked to hold a tool and find food," Foerder said. "You wouldn't be able to do it."

The most intriguing discoveries about elephant minds have come not in captive studies, however, but from the field. Just by watching Amboseli's families, for example, McComb, Moss, and Njiraini know that elephants are extremely adept tool users, on a par with chimpanzees and crows. They pluck branches to scratch ticks off their bodies, pick up logs and rocks to throw at adversaries, and drop logs onto electric fences to disable them. Asian elephants have also been seen making tools: when in need of a fly whisk, they break off a tree branch and shorten it, until they have a tool of just the right size to keep the flies away.

Elephants are also vocal learners; they're one of those few species that learns calls by listening to and imitating others—or whatever is available. The pachyderms' vocal-learning ability was reported only in 2005, after Joyce Poole, another member of the Amboseli elephant project, discovered that a ten-year-old orphaned female living in semicaptivity near a major highway in Kenya's Tsavo National Park imitated the sounds of the trucks that passed by in the night. Other project researchers have shown that the Amboseli elephants are as keen with their trunks as they are with their ears. They understand that the odor of Masai men means there is danger nearby (the Masai spear them at times to protest the national park's policies, which, they say, favor the elephants over their cattle); and they can keep track of the location of their family members by smelling where each one urinated.

From long-term studies, elephants have proven themselves to be among the best social networkers in the animal kingdom—a cognitively demanding skill, as most of us in this social media age well know. Elephant society is centered on the family, which consists of the matriarch, her daughters, and their descendants. Often a family will also include the matriarch's sisters and their offspring. There may be a few young males, but after they reach puberty, around age fourteen, they spend time with young males in other families, where they learn the ways of being a bull. Older adult males alternate between sexually active and inactive periods; when they're inactive they join all-male groups, and when sexually active, they move from family to family in search of females in estrus.

Related families like to hang out together, forming what Moss and Poole call bond groups. When related bond groups join together, the total aggregation is termed a clan; a clan can include several hundred elephants.

These aren't random gatherings, McComb told me as we maneuvered into position for the day's experiment. Moss had first wanted her to test the idea—based on her observations of which elephants spent time together—that members in a clan know one another. It even seemed that they could distinguish the contact calls—deep, rumbling reverberations—of close friends and family members from the calls of elephants they'd met only casually, or ones they'd never met at all.

"And the answer is yes, they can tell the difference between their allies and those they rarely meet," McComb said.

But as they analyzed their data, McComb and Moss realized that the elephants were not equally adept at distinguishing the calls. The oldest matriarchs were far better than young, thirty-year-old leaders.

In a related study, the team also discovered that the older matriarchs were the best at deciphering recordings of lions' roars. Every elephant matriarch could tell the difference in the number of lions in a pride: they knew when there were three lions roaring versus one, and were more cautious when they heard three. But again, only the oldest matriarchs could correctly identify the most serious threat—male lions—and focus their defense on them. Even when alone, a male lion can kill an elephant calf. The difference between a male lion's roar and that of a female's is "very subtle," McComb said. "It's very difficult for us to tell them apart." When a matriarch hears male lions roaring, she alerts the rest of her family, sending a signal that brings the family together in a tight formation the researchers call "bunching."

"In this situation, she may go right to the front, with the other older females, and lead her family forward, in search of the lions—or she'll move to the rear to steady everyone," McComb said. "She does not want to be ambushed; she wants to either prepare for an attack by bunching her group together or confront the lions and chase them away."

McComb's new test was designed to further delve into the matriarchs' ability to make decisions—in particular, what course of action they choose in situations involving other socially dominant females. She would play the contact call of an older, unknown female elephant to the family we were following and watch their reactions.

As McComb had hoped, the herd stopped to feed on the fresh grass of the mudflats. Shannon stopped about three hundred feet from the largest females and switched off the engine. Some of the younger elephants looked over at us but then resumed their feeding.

"Who do we have here?" McComb said, turning to Njiraini.

"It's two families, the OAs and the OA2s," Njiraini said. She flipped quickly through a thick card file containing photos and sketches of the Amboseli elephants—some 1,700 animals—showing their distinguishing features, the frayed and nicked ears, scars, and other marks that made it possible to identify each elephant. Each card also contained a condensed biography of the individual elephant—its parentage and age, number of offspring, known injuries, and usual companions.

"You see the one with the long, upturned tusks?" Njiraini asked. "That is Olympia, the matriarch of the OAs. And that other big female is her aunt, Orabel, the leader of the OA2s."

Njiraini called out some of the other, younger elephants' names: Oasis, Onyx, Oralee, and OmoR (for Omo River). Years ago, when she started the project, Moss decided to name the elephants, believing it would be easier to remember names rather than numbers and that—as a human—she would feel more connected to the animals she studied. She didn't give them random names but devised a method that could be easily entered into a computer program. The first family she photographed she called the AA family. She gave each member of that family a name that began with the letter A: Alison, Amelia, Agatha. The first three letters of each elephant's name became their computer codes: ALI, AME, AGA. When she identified a second family, she called it the BB family, and so on.

Moss and her team have identified sixty-one elephant families and have run through the alphabet twice to give each a different code. And because families sometimes split apart into new, separate groups, she has also added numbers. That was the case with the OAs and the OA2s, Njiraini explained. A few years ago, they were one family—the OA family, led by the matriarch, Orlanda—until Orabel and her calves began to spend more time on their own. Then Orlanda perished in the drought.

When a matriarch dies, the family goes through a period of upheaval and confusion. Sometimes families break apart. Other times, they join closely related bond groups, as the OAs appeared to be doing. In nearly every case, the oldest female becomes the new leader—although how the elephants know who is the oldest in their group remains a mystery, McComb said. "That's something else we'd like to get at. I'm trying to think up an experiment."

In the OAs' case, the oldest female now was the relatively young, thirty-year-old Olympia. "Maybe she's a little uncertain," Njiraini said. "I think that's why she's come back to see her aunt, Orabel, who's forty-six years old. Maybe they'll all join up in one group again."

McComb studied the OAs through her binoculars. Two young, three- to four-year-old calves stood close to the two matriarchs. There weren't any younger calves or newborns, which made it a good group for the experiment. Other families we'd seen that morning did have tiny calves—some as young as one or two weeks old. At that age, their ears are pink, their skin a dark, glossy gray, and their trunks loose and wobbly. They haven't mastered control of that appendage, Njiraini had explained, so it dangles in front of a young calf's face like a long noodle. "Newborns can look so delightful," McComb had said when one wide-eyed tyke darted out from between its mother's legs and flapped its rosy ears—as if trying out how they worked.

The calves with the OAs were frisky, too, playfully twining their trunks. They had sprouted tiny tusks, which poked out from their upper jaw like a kid's new front teeth. No longer nursing, they nevertheless didn't venture far from their mothers, whose massive legs formed a kind of protective crib around them.

"It looks like they're all relaxed," McComb said. "So, let's do it."

She pulled out her recorder and a clipboard with a data sheet, and Shannon climbed onto the roof of the Land Rover to film the proceedings. When he signaled that he was ready, McComb switched on the recorder, which was connected to a large speaker strapped in the back of the Land Rover. Instantly, the air filled with an elephant's deep-throated, thundering call. The speaker was directly behind Njiraini and me, and our seat shook as the elephant's vibrato rolled past us. This was the slightly altered contact call of Abby, a matriarch who lives in South Africa; the team had acoustically manipulated her call to make it sound like a forty-five-year-old female, and so they referred to the recording as Abby-45. Elephants make contact calls after they've been feeding and have become separated from their families, saying in essence, "I am here. Where are you?" The recording lasted only five seconds so that the elephants would not figure out exactly where it came from. "We don't want them to think it came from us," McComb said.

Although the call was brief, many of the elephants stopped feeding. The voice, after all, sounded not like a shy creature, but like a confident, dominant, assertive individual—even someone as inexperienced with elephant calls as I was had understood that. Most of the elephants had been standing parallel to the car, and the two matriarchs maintained that pose. But some of the younger elephants moved cautiously toward their leaders, and the two calves stayed close at their mothers' sides.

"Orabel is listening," McComb said. "See how she's moving her head slightly from side to side, and how she's holding her ears out? And Onyx is listening. Now Orabel and Olympia are down-trunk smelling." They used their trunks like inverted, odor-detecting periscopes, holding them close to the ground, while lifting the tips to capture any scent of this strange female. With their trunks on the ground, they could also pick up vibrations from any more calls the invader might make.

Elephants always listen and smell first, using their keenest senses to spot an intruder like Abby or a predator, McComb had told me earlier. Only after finding her aurally or nasally would they look for her visually.

"Now Olympia is moving behind Orabel," Njiraini said. "And Orabel is streaming. See that dark mark on the side of her face?"

Elephants have temporal glands between their eyes and ears. When an elephant is excited or worried, a secretion flows from the glands, leaving dark trails down the sides of its face.

The young Onyx moved closer to Orabel, and the two exchanged greetings, but their calls were far softer than the loud, domineering voice of Abby-45.

"Oh, Onyx is a little worried," Njiraini said.

"Yes, but look how Orabel is responding," McComb said. "She's staying in one place, listening, and keeping everything calm. We don't know exactly how it works, whether she's giving them an olfactory or acoustic clue, but that's what she needs to do—keep all her family members steady. She holds the line. If she gets agitated, then the others will get even more upset, and the whole thing can ratchet out of control."

A strange dominant matriarch, like Abby-45, is a threat. She could be encroaching with her family, setting off a competition between the two groups.

Cynthia Moss has witnessed frightening encounters between nonallied families. When the EBs and WBs gathered in the same area to feed, one of Moss's favorite females, Echo, marched up to the WBs' new baby, scooped it up with her tusks, and pitched it through the air. The calf landed about three feet away, unhurt. Other members of the EBs also attacked, always taking out their aggression on the baby, perhaps in an attempt to get the WBs to leave—although the EBs' tactics didn't work. The WBs stood their ground. A few days later, the EBs and Echo were themselves attacked when Freda, the matriarch of the FBs, hustled up to Echo's baby, Ely, and kidnapped him. Then the FBs hurried off across the plain, keeping Ely surrounded. Echo and her friends pursued and eventually managed to reclaim little Ely. In both cases, Moss thinks the matriarchs were using the calves to settle disputes about the families' social rankings: which family gets to eat here and which one can be pushed away.

If that's how matriarchs from families that know each other can behave, who knows what a foreign female like Abby-45 would do? This is what Orabel and Olympia had to determine. And it was why they and their families were worried.

Slowly—indeed, so very slowly that unless McComb had pointed it out, I wouldn't have noticed that this was a reaction at all—the other elephants drew closer together, clustering around their youngest members. It was the elephants' defensive "bunching" response. But Orabel didn't change her position. She stood between us and the members of her group like a wall. Calmly, she pulled up a hunk of grass and tucked it into her mouth. Following her cue, the other elephants settled down to feed as well. All of this was a sign that Orabel didn't want to waste too much of her family's time on this vocal intruder. Only the dark trickle down Orabel's temples and the angle of her ears revealed that she was even slightly concerned. Olympia did not seem too rattled, either. She was obviously listening, her ears still flared slightly, but she kept eating.

"We want to understand how these older, wiser matriarchs decide whether to move or to stay calm. What are they listening to in a call? How does it inform decisions?" McComb said.

An inexperienced matriarch might turn to pull her family in tightly around her, stop them from feeding, and get them all to listen and smell. If it was a false alarm, she would have wasted their time and energy. Too many poor decisions would affect her family's reproductive success. "That's not a speculation," McComb said. Families with poor leaders are known to have fewer calves—they lose in the genetic sweepstakes of life.

We sat watching the OAs and OA2s for another half hour. They all continued feeding, munching on grass, while white egrets settled on their backs and small flocks of songbirds flitted among the bushes. To my eyes, they looked as fully relaxed as when we first joined them. But McComb and Njiraini continued to point out subtle clues—the movement of an ear, the twitch of a trunk—that indicated the elders had not forgotten Abby-45's brassy call.

In some tests, matriarchs have actually led their families in search of the invader, streaming right past the team's Land Rover. Orabel made a different decision. About thirty-five minutes after hearing the call, she took a few steps in our direction, seemingly because she wanted to feed on the next patch of green grass. She uprooted a bunch, stuffed it in her mouth, and kept going, moving slowly but purposefully past us. The other elephants fell in line, some of them shoulder to shoulder. And as each one came close to the Land Rover, she lifted her trunk, still checking for the scent of that stranger.

"They're on the lookout," McComb said. "They're *thinking* about that stranger, but they trust their matriarch, so they're not out of control. How do the matriarchs do that? How do they keep everyone calm while deciding what to do in a situation like this? We're trying to get to that deeper level of what's going on inside their minds." McComb thought that after they'd tested enough families with Abby's and other elephants' calls, and reviewed their data and videotapes, they might spot the matriarch's moves that signaled "Stay calm, no need to worry" to her family.

"These kinds of discoveries really do matter," McComb continued. "If most of the herds are led by young, inexperienced matriarchs because their elders have died from droughts or poaching, then their families are at risk. They may not be as successful; they may have fewer offspring."

McComb didn't have to tell me what that meant. Elephant populations have plummeted precipitously in the past thirty years—from an estimated 1.3 million in 1979 to 450,000 in 2007. Many conservationists are deeply worried that elephants may become extinct in this century. Fewer offspring because of a young matriarch's inexperience and poor decision making will only hasten their end.

Most mammals are born with brains that don't expand much beyond birth; their brains are about 90 percent of their adult weight. Human infants, in contrast, are only about 23 percent of their final capacity—a difference that neuroanatomists explain by our need to learn. Baby elephants, too, have much to learn, and their brain size at birth is around 35 percent of what it will be at maturity. When poachers target the matriarchs or older females—as they often do, because older elephants usually have larger tusks—they also destroy that lifetime of learning and knowledge. For an elephant family, the death of a matriarch must feel like losing an encyclopedia, or an entire library—and for us, the loss makes stopping the poaching even more urgent, if only to protect the experienced matriarchs, who keep their families out of harm's way.

WHEN MCCOMB WASN'T TESTING the elephants, she was watching them. She kept a close eye on each family's dynamics, looking for situations that might inspire an experiment. Sometimes, she told me, she just couldn't figure out a way to test what she wanted to know—so she thought up another way to get the answer.

One afternoon, as we followed the BB family, we were suddenly engulfed in an exuberant celebration of trumpets, screams, and rumbles as another family came trundling up to greet them. Trunks were extended to their full length and intertwined, like clasping hands; some females spun in circles, while urinating and defecating; they all lifted their heads high, rapidly flapped their ears, and roared some more. We laughed aloud at their joy.

Before this raucous greeting ceremony, the BBs had stopped feeding to listen to the approaching calls of their friends. Or so it seemed. "It always looks like they know who is coming," McComb said. "And when you play the call of a missing family member to elephants, they run trumpeting toward the speaker. It looks like they expect to meet that individual; that they have someone 'in mind' they expect to see."

We do this all the time, most often by matching auditory cues with a visual memory, such as when talking with a friend on the phone. The sound of our friend's voice triggers a mental image of the person. It's called "cross-modal perception"—meaning that we're matching information we perceive with one sense (in this case, our hearing) with information we've previously gathered via other senses and stored in our memories.

Elephants and many other species from vervet monkeys to songbirds and parrots seem to be doing something similar to what people do. But as Mc-Comb well knows, showing with hard data that an animal actually recognizes a specific individual isn't easy to do.

When she became stumped about how to set up such a test for elephants in the wild, she turned to domesticated horses back home instead. "It was easier to devise a test where you had some physical control of the animals," she said.

McComb, together with her colleague David Reby and graduate student Leanne Proops, devised an experiment that could be done in a stable—which is a natural environment, she notes, for domesticated horses. A groom would hold one horse loosely by its halter rope while a second horse from the same herd was led past and guided to a position behind a barrier, where it remained hidden from the first horse. A few seconds later, the researchers, who were also behind the barrier, played a recording of a whinny. Sometimes they used the whinny of the horse that had just walked by; other times, they used the whinny of a different member of the herd. For instance, in one variation of the test, Silver watched Pepsi walk by and disappear behind the barrier. He might then hear Pepsi's whinny, or that of Fi, a horse Silver also knew. In both situations, the scientists filmed Silver's reactions. When Silver heard Pepsi's whinny, he looked briefly in the direction of the call but then quickly resumed what he had been doing. What he heard matched what he had just seen. But when Silver heard the call of Fi, he immediately turned to look at the barrier and gazed at it for some time. The scientists gave this test to several horses and always elicited the same surprised response when the horse heard someone different from the one it had seen.

"It shows the horse had an expectation," McComb said. "It expected to hear the whinny of the individual that had just walked by, not someone else.

It means that a horse has pictures in its mind of the horses it knows. I'm sure many other animals do this as well."

The idea that animals think in pictures had been suggested before, though not with such clear evidence that they do. McComb's, Reby's, and Proop's discovery—published in 2009—was hailed as a tremendous breakthrough in animal cognition research.

"It gives us something concrete," said McComb. "We just need to work out ways to reveal the other pictures in the minds of animals. That would give us a better idea of what their minds are like, or at least the basic building blocks they use to represent the world. It seems they're doing something very similar to what we do by representing someone they know with a mental picture. But how does it work in animals? They don't have language, so they probably don't have a mental narrator as we do. Maybe it's more like a silent movie, or like a meditation."

Given their intensely social natures, elephants likely retain their memories for many years as well, McComb thinks. "We've just touched on this," she said, but one of her experiments is very suggestive. She played the contact call of a female who'd left her family twelve years before to join another; the female's original family members gave their own throaty calls in reply to the recording. Elephants also have the necessary neural anatomy for long-term memories—their brains have especially large and complex frontal lobes, which are important for storing and retrieving memories of scent, touch, smell, and sound.

There's little doubt that elephants have prodigious memories. Randall Moore, an elephant trainer who helped return Owala, an American zoo elephant, to the wild in Pilanesberg National Park, South Africa, tells a revealing story about hers. One day, twelve years after her release, Owala was bitten by a hippo. The park veterinarian needed to treat her wound regularly, cleaning it out and applying ointments. He didn't want to have to immobilize her for every treatment, so he asked Moore if he could assist. Moore had not seen Owala since he set her free, but he agreed to help; he traveled to the park and began calling out her name. Owala walked directly out of the bush to meet him, raising her trunk in greeting. Then, following Moore's commands, she calmly lifted her foot and let the veterinarian treat her.

And, too, there was another experiment McComb staged, using the call of a fifteen-year-old female elephant who had died. She played the deceased elephant's call to her family twice, once three months after her death and again twenty-three months later. They rumbled back to her in greeting, and walked directly to the loudspeaker. "They hadn't forgotten her," McComb said, "but I was uneasy doing that test." It may have left the elephants confused or raised some feelings in them akin to sorrow.

EVERY ELEPHANT RESEARCHER has witnessed elephants apparently mourning and grieving; their tales of elephant grief are among the most affecting stories in the literature of animal behavior.

Longtime elephant watchers Cynthia Moss, Joyce Poole, and Iain Douglas-Hamilton agree that elephants have some basic concept of death. Just as we recognize the body of a dead human or a human skeleton, elephants recognize the carcasses and skeletons of their kind. They smell the bones of their dead, even old ones bleached by the sun, and caress them with their trunks. Often, when an elephant has just died, other elephants will back up to touch its carcass gently with their hind feet, then cover the body with dirt and sticks, and stand guard. (Intriguingly, elephants have done the same to the bodies of people that they either find dead or have killed. One young orphaned elephant in a South African sanctuary shrieked and moaned when it discovered the buried remains of its daily companion, a rhinoceros, that poachers had killed for its horn.) Chimpanzees, gorillas, some corvids, and dolphins also spend time with their dead, but overall, most species do not.*

Elephants also often struggle to help dying companions, revealing their empathy and compassion. In 2003, Douglas-Hamilton's team came upon Eleanor, a matriarch in Kenya's Samburu National Park, as she was dying. When Eleanor fell to the ground, Grace, a matriarch from another family, used her tusks to lift Eleanor and help her to her feet. Eleanor fell again.

* Marc Bekoff, an animal cognition scientist, has seen a male fox near his home kicking dirt and sticks over the carcass of his mate, apparently to bury it.

Grace appeared alarmed by this and called out, then tried again to help Eleanor to stand. Grace stayed by Eleanor's side, even when Grace's own family moved on. Eleanor died that night. Over the next several days, her family members and other elephants spent time with her body, nudging it, and feeling and smelling it with their trunks. Her six-month-old calf never left its mother's side, even after park rangers cut out her tusks to make sure they did not fall into the hands of poachers. In photographs, the calf stands like a tiny sentinel next to its mother's body, while the rest of Eleanor's family, bunched close together, looks on. Her young calf disappeared three months later and probably was killed by a predator.

The pattern of behavior when elephants encounter their dead rarely varies, Moss says. "They stop and become quiet and tense in a different way from anything I have seen in other situations. First they reach their trunks toward the body to smell it, and then they approach slowly and cautiously and begin to touch the bones, sometimes lifting them and turning them with their feet and trunks. They seem particularly interested in the head and tusks. They run their trunk tips along the tusks and lower jaw and feel in all the crevices and hollows in the skull. I would guess they are trying to recognize the individual."

In some situations, elephants will move a skeleton's bones, carrying them with their trunks for some distance before dropping them. "It is a haunting and touching sight," Moss says, "and I have no idea why they do it."

Whenever the Amboseli researchers find an elephant's remains, they collect the lower jaw so that they can determine the elephant's age. They leave these jawbones near their camp, and without fail, passing elephants pause to inspect the remains. One time, a few weeks after they brought back the jawbone of an adult female elephant, her family came through the camp. Every member stopped to inspect her jawbone and teeth, but one individual—her seven-year-old son—stayed behind, feeling and stroking the jaw. He turned it over with his foot and trunk and smelled it repeatedly. "I felt sure that he recognized it as his mother's," Moss wrote.

Behaviors like these are strange and compelling—perhaps because they suggest that an elephant is reflecting on someone it cared for and may have loved. It may be why grief in animals strikes us, even more than other forms of cognition, as evidence of a mind—an awareness of oneself and one's feelings for others. As Joyce Poole wrote after watching an elephant investigate her mother's bones: "Why would an elephant stand in silence, over the bones of its relative for an hour if it were not having some thoughts, *conscious* thoughts, and perhaps memories?"

It is tempting to imagine that this is the case, but are we merely attributing human thoughts to elephants?

"No, something is definitely going on there," McComb said. We'd completed the day's experiments and returned to the research camp. There, set out in long rows in a grassy clearing, were the jawbones and several skulls of the twenty matriarchs who had died in the drought. These were the remains of some of Amboseli's most famous elephants: Grace and Isis, Leticia, Lucia, Odile, and Ulla, Freda, Xenia, and Orlanda. Even Echo, known and loved around the world, had succumbed to drought-related malnutrition. She had died about a mile from Cynthia Moss's tented camp. Her sister, Ella, is now the matriarch of the EBs and has been seen stopping by to visit Echo's remains.

Is it possible that elephants recognize their dead from a lingering scent?

"We tried to answer that question ten years ago," McComb said, "but the results weren't conclusive," perhaps because the elephant skulls they used were too old. She and Moss placed a trio of animal skulls—one buffalo, one rhinoceros, and one elephant—close to several elephant families. The elephants always stopped to inspect the elephant skull but largely ignored those of the other species. The elephants also picked up and smelled a piece of ivory the scientists placed near them but not a hunk of wood. "They really are interested in elephant bones. There's no question about that," McComb said.

Next, McComb and Moss gave the elephants a test to see if they preferred the skull of their own deceased matriarch over the skulls of two other unrelated matriarchs.

"That was the 'big' test, our way of trying to see if they remembered a particular individual," McComb said. "But they didn't." They were simply interested in elephant bones and ivory, period.

"Which is still fascinating," McComb said. "They really do regard the bones of their own species as different from other bones. Why is that? It does seem that their social connection with other elephants is so strong it goes beyond death."

Might their behavior be a kind of ritual, a mourning for elephants in general? I wondered.

McComb shook her head. "We just don't know." She had not given up, though. To her, it was like a cold case, and she wanted to take another stab at cracking it. Thinking that maybe the teeth in the lower jawbones retained more of the elephant's scent, she and Moss planned to present three jawbones from the recently deceased matriarchs to a specific family. One jawbone would be from the family's old matriarch. Would they touch it more than they did the other two? If so, could that be construed as evidence that they remembered her and missed her? That they grieved for her?

McComb didn't want to guess. "The experiment will tell us something more. It should give us another clue.

"The ultimate thing," she added, "would be to test if they remember ivory from a family member." Such an experiment was not about to happen, though. The park authorities always collect the ivory from dead elephants, storing it in locked warehouses to keep it out of the hands of poachers.

IMAGINE THAT YOU'RE A FOUR-YEAR-OLD male elephant calf, following your mother across a grassy plain. Walking nearby, sampling the grasses as they go, are your close relatives—your five-year-old sister, a dozen of your aunts, and all their kids. It's a quiet morning—until the air erupts with the sound of a helicopter, flying low, circling above you and your family. Your mother begins to run, but the helicopter's noise and the dust confuse everyone. And then the shooting begins. Around you, your family is dying—your mother, your sister, your aunts. Everyone is killed in a few minutes—except you and a young female cousin. You huddle next to your mother's remains, and when the people come, they chain you to her carcass for several hours. Later, they load you and your screaming cousin into a truck, and when you emerge, you're in a new land, Pilanesberg National Park, South Africa. Here, there are other youngsters, a year or so older than you, and a couple of nice old female elephants, who take you under their protection. But there are no adult male elephants. How do you and your cousin grow up? What kind of elephants do you become?

This is another kind of experiment—although when it was started the wildlife managers didn't think of their activities as experimental. Culling operations, such as this imagined one (although based on actual culls), were part of a plan to manage the number of elephants living corralled in fenced reserves.

In the 1960s, South African wildlife officials worried that if there were too many elephants in a fenced park, like Kruger, they would ultimately eat their way through the vegetation, destroying their habitat. To control the elephants' population, Kruger park managers authorized the killing of entire elephant families. Young, weaned elephants, those around four to ten years old, were often spared and shipped to other reserves that either lacked elephants altogether or had smaller populations. Some of these youngsters ended up in Pilanesberg.

By 1990, there were at least seventy junior elephants in Pilanesberg, under the supervision of the two matriarchs, and some of those "juniors" had grown into teenagers. Under normal circumstances, teenage males would have left their natal families to join a herd of older bulls. But these weren't normal circumstances, and the young males wandered off to form their own groups. In 1993, two of the teenagers bashed a tourist's car, chased him from it, and killed him. Park officials killed these two young elephants, labeling them "rogues." A few months later, in March 1994, game scouts reported that an elephant had driven its tusks through a rhinoceros, killing it. Authorities were skeptical, but two months later, fifteen more rhino carcasses were discovered. All had deep wounds from being tusked and trampled.

Teenage elephants in similar reserves were also behaving aggressively and killed five people.

What was wrong with these youngsters? And other than dismissing the elephants as "rogues" and killing them, could wildlife biologists do something to resolve the problem?

"That's where long-term studies like Cynthia Moss's and Joyce Poole's really do help," said McComb the next day. We were off again with Shannon at the wheel of the Land Rover in search of another herd to test with the voice of Abby. But for this test, McComb would be playing a different version of Abby's voice, one that made her sound like a fifteen-year-old. Would elephants perceive the teenager as a threat? McComb didn't think so, but cautioned, "You never know until you do the test. Teenagers can cause trouble, as we know from Pilanesberg."

Especially unsupervised teenagers. When Joyce Poole heard about the Pilanesberg elephants killing people and rhinos, she suspected that the problem was linked to the young elephants' abnormal society. Poole's early specialty was the male African elephant. She discovered that males of the African savanna species, *Loxodonta africana,* experience periods of *musth,* just as males of the Indian species, *Elephas maximus,* do. Musth is similar to the male rutting, or mating, period that numerous hooved animals, such as antelope and elk, undergo. It's a time when male hormones surge, making males especially aggressive, which can help them outcompete other bulls for access to females. Although males are physiologically capable of breeding in their late teens, they don't start coming into musth until their late twenties.

Although Poole doesn't know exactly how it works, the older males seem to suppress the musth periods of younger males, which helps the younger elephants control the aggression and rage they may feel from their surging hormones. Poole and other elephant researchers suggested that the problems at Pilanesberg were due to the lack of older bull elephants. In March 1998, six large bulls from Kruger were moved into the park—and the problems with the young teenaged male elephants stopped. There have been no more reports of elephants killing rhinos in Pilanesberg, and South Africa has not culled any more elephants since 1994 (although the country maintains its right to do so).*

But many questions remain about the Pilanesberg elephants, says McComb, who is testing them with the same experiments she gives to the Amboseli elephants. "Most of the Pilanesberg elephants are young, with young matriarchs as their leaders. They don't have the expertise of the matriarchs here, and it shows in how they make decisions. They get rattled very easily.

* South Africa enacted a ban on culling in 1994 but lifted the moratorium in 2008.

"Elephants just have so much to learn," McComb continued. "I don't specifically focus on the similarities between animals and ourselves. I find the differences just as interesting. But elephants are very like us in having a long youth and learning period. And when that is disrupted, they suffer."

Human psychologists know that a break in the mother-child bond in people can cause a reduced capacity for empathy and an increased tendency for violence. We humans, too, often lose our moral bearings if our social structure is disrupted.

Some researchers, such as the psychologists Gay Bradshaw and Allan Schore, think that the Pilanesberg elephants and others that have experienced the social trauma of culling are similarly psychologically damaged and may be suffering from something like post-traumatic stress disorder. The elephants, say Bradshaw and Schore, show the same symptoms that people with PTSD do: they're depressed, easily startled, unpredictable, and violent.

All of which shows how much "young elephants need their families," said McComb. They need the rules and guidelines of their society and the wisdom of their matriarchs.

Perhaps this is something that elephants know, too. For when McComb played the recorded contact call of the teenaged Abby to the AA family, the twenty-three elephants reacted quickly, bunching together and turning to face us, all ears in the listening pose, all huge bodies frozen in place. Allison, the matriarch, lifted her trunk high in the air as heavy secretions spilled from her temporal glands. Other elephants mimicked her behavior and hiked up their trunks.

"Maybe it's the novelty of the call," McComb said. "They aren't sure what to make of it."

Something about that call was wrong, the elephants seemed to say. They had reacted so quickly and strongly that even McComb was surprised. "It could be that they don't expect a young female to make such a loud call," she said. "When they react so quickly to something, it seems to be almost an intuitive response, like when we make decisions unconsciously. Or maybe they were spooked earlier today by something else." There were many possible explanations for the elephants' response, which is why the scientists needed to carry out the experiments multiple times—only then would consistent patterns of behavior emerge.

McComb gave a wistful sigh. "I would love to know what they're thinking right now, how they're experiencing this. That's the one thing I'm sure of: we don't have a monopoly on experiencing the world, on having events mean something."

Immediately after a culling operation, observers have noted, other elephant families come to investigate the scene—even though park rangers always thoroughly clean up the area and leave no bodies or bones behind. As "clean" as the grounds of a cull site may appear to us, they still attract elephants, who often travel a day or more to reach the spot. Do they know that something terrible has happened because the elephants under attack sent out low, infrasonic rumbles of fear and distress? Such calls can travel two miles or more, and that may be how elephants find their way to the culling grounds. After investigating and smelling the earth, though, the visiting elephants don't return. Even if it is excellent habitat for elephants, the land remains empty of elephants for years—perhaps contaminated by odors of terror, fear, and death.

McComb wouldn't venture a guess about how elephants might mentally represent the idea of "sorrow" or the concept of an action or place as being "wrong." These are human words, and perhaps they are not the right ones to attach to elephants rushing to the calls of distress on a distant culling ground. All one can say is that it would make sense for those calls, the smells, that bloodied ground to linger in the mind of a wise matriarch.

ONE OF THE MORE STRIKING DISCOVERIES in neuroscience in recent years is the finding that elephants, whales, great apes, and humans all possess a peculiar kind of brain cell. These neurons were first discovered in human brains in the nineteenth century and were named von Economo cells after the Romanian anatomist Constantin von Economo, who identified them. At first, these spindle-shaped neurons were touted as the cells that "make us human," because they're connected to our feelings of empathy, love, emotional suffering, and sociality. Then, in 1999, two other researchers, Patrick Hof and John Allman, spotted von Economo cells in the brains of all the great apes; others recently found them in monkeys. Allman has searched without luck for the cells in more than a hundred other species, from sloths to platypuses. So it was big news when, in 2007, he discovered spindle cells in the brains of whales, dolphins, and elephants. But it was a puzzling discovery, too. Why should such a disparate group of animals have these specialized cells?

From an evolutionary point of view, it's not surprising that primates and humans have von Economo cells, since we are in the same lineage. But primates and humans haven't shared an ancestor with whales or elephants since about the beginning of the mammalian lineage, some sixty million years ago. It seems that cetaceans and elephants evolved their spindle cells independently. What factors would produce such emotionally specialized brain cells?

Allman thinks part of the answer lies in the size of the animals' brains—most species that have spindle cells also have notably large brains—and in the location of the cells. Von Economo cells are always found in two regions of the cortex associated with emotionally charged, visceral judgments, such as deciding whether a fellow animal is suffering. And part of the answer lies in the size of the spindle cells. They are unusually large, enabling them to act like high-speed circuits, fast-tracking information to and from other parts of the brain, while bypassing unnecessary connections. These are the kind of cells, Allman argues, that would be especially useful to an animal living in a complex society—a society in which making accurate, intuitive decisions about another's actions (or facial or vocal expressions) is crucial for your family's and your survival.

Allman has the luxury of being able to monitor human brains in the process of making decisions. So he's seen spindle cells activate when a mother hears a baby cry. McComb can't hook up the elephant matriarchs to brain-scanning machines. But the matriarchs' behaviors suggest that their spindle cells, too, are firing furiously during McComb's experiments. Allman also suggests that these cells underlie the empathetic behavior elephant researchers have recorded, such as when two matriarchs saved a drowning elephant calf in Amboseli, or when another elephant brought water to a dying companion. And if that is the case, then it's most likely that these spindle cells kicked in, too, during those times when the Kruger elephants visited the culling sites.

"It's the one thing I have to constantly remind myself of—how different we are from elephants," McComb said, "because in other ways, we can seem so much alike."

Many people—scientists and nonscientists alike—have the same feelings about whales and dolphins. Some researchers have gone so far as to label dolphins "humans of the sea" and to call for them to be recognized as persons. Yet not only do whales and dolphins have very different body architecture from ours, but they live in a completely different, almost hostile, environment—at least for us.

Why do dolphins strike us as such kindred creatures?

7

THE EDUCATED DOLPHIN

> Man had always assumed that he was more intelligent than dolphins because he had achieved so much—the wheel, New York, wars, and so on—while all the dolphins had ever done was muck about in the water having a good time. But conversely, the dolphins had always believed that they were far more intelligent than man—for precisely the same reasons.
>
> DOUGLAS ADAMS

It was the start of the dolphins' day at the National Aquarium in Baltimore, Maryland, and five dolphins were lined up at the edge of a large pool. Their big, smooth heads poked out of the water, and they squeaked and clicked at their trainers, five young women dressed alike in blue T-shirts and khaki shorts. Shiny buckets brimming with silvery fish sat next to the trainers' feet. Like coaches, the trainers talked, signaled, and whistled to the dolphins, giving directions. When the dolphins did something right, the women cheered and tossed fish from the buckets into their charges' clacking jaws.

One dolphin leaped in the air, her body arced as gracefully as a ballerina's. Another did a tail stand and hopped backward over the pool's blue surface, as if performing an aquatic version of the moonwalk dance step. A third one rolled onto her back close to the pool's edge, exposing her genital and anal slit while her trainer kneeled down and, with a gloved hand and plastic scoop, took a fecal sample.

There were tail slaps, huge, spinning jumps, and even bigger splashes—all part of the dolphins' exercise and training routine. I watched the proceedings with Diana Reiss, a comparative psychologist and animal cognition researcher from Hunter College in New York City. Reiss, a petite brunette with a warmly busy manner, was there to begin a new experiment with young dolphins. It was part of her research using mirror self-recognition (MSR) tests to upend old ideas about which animals are able to recognize their reflected images—a skill that scientists have long linked to self-awareness. Like the elephants in the Bronx Zoo, animals that pass the MSR test are thought to belong to an exclusive club, one whose members can think about themselves and their actions. (Reiss was a member of the team that tested the Bronx elephants for this ability.)

Gordon Gallup Jr., then a psychologist at Tulane University in New Orleans (and now at the State University of New York at Albany), invented the mirror test in 1970 after reading Darwin's account of watching an orangutan, Jenny, at the London Zoo in 1838. Darwin noted that when the young ape saw her reflection in a mirror, she appeared to be "astonished beyond measure." Although Gordon doubted that animals had minds, he wondered if Jenny had recognized herself in the mirror. He then came up with a method for testing apes for this ability and gave his exam to four chimpanzees, two young females and males, who all passed. Gallup, the previous animal-mind skeptic, concluded that the chimpanzees did understand who the chimpanzee in the mirror was; they realized that their reflections were external representations of themselves, and they used the mirror as a tool to study and watch themselves. It was the "first experimental demonstration of a self-concept in a subhuman form," Gallup wrote in *Science*. He also tested three species of monkeys, who all failed; they continued to try to relate socially to the image in the mirror, treating their reflection as another monkey. "Our data suggest that we may have found a qualitative psychological difference among primates," Gallup wrote, "and that the capacity for self-recognition may not extend below man and the great apes."

Since then, Gallup's MSR test has come to be regarded as the gold standard in determining if species other than chimpanzees and humans are self-aware. The test follows strict rules. An animal is allowed to spend some time with a mirror to get used to the mirror's properties. Later, the creature is daubed with a colored but odorless mark and is given a mirror again. Species that pass the test use the mirror to look and pick at the mark, just as you or I would scrape at a blob of mustard we notice on our lapel when looking in a mirror. For thirty-one years, only humans and our closest relatives—chimpanzees, bonobos, orangutans, and gorillas—successfully passed the MSR exam. Many scientists seized on those results, as Gallup had, as evidence of a major divide between the human-ape lineage and all other species. Something special was believed to have had happened in our evolution to make us self-aware, conscious of our actions and intentions.

In 2001, Reiss and Lori Marino, a neuroscientist at Emory University, shattered the idea that only humans and great apes could pass an MSR exam, after testing two adult dolphins, Presley and Tab, at the New York Aquarium in Brooklyn. Even without fingers to explore their marks, Presley and Tab passed, the two researchers reported in a paper they published in the *Proceedings of the National Academy of Sciences*. Presley, in particular, reacted strongly to the black triangle Reiss had drawn on his right pectoral fin. He flipped head over tail a dozen times in front of the mirror to get a glimpse of that mark.

"It was very purposeful behavior," Reiss said. "They wanted to look at those marks, and they positioned themselves to get a closer view," actions that indicated they knew they were looking at themselves.

Although not all scientists accepted the results (because the dolphins couldn't pick at their marks), Reiss and Marino's study was still widely hailed as a major breakthrough. Theirs was the first study to bridge the self-awareness chasm, making it possible to begin considering the evolutionary roots of this ability. As with other mental abilities, it seems more likely that there is a continuum to self-knowledge rather than a sharp divide between species, a notion that received even more support when Asian elephants and European magpies (a member of the brainy corvid family, which includes crows and jays) also passed the mirror test. One elephant used its trunk to touch the mark, and the magpies used their beaks to peck at yellow dots that had been affixed to their plumage. Animal cognition researchers Marc Bekoff and Paul

Sherman have suggested that many more species might qualify as being self-aware if researchers devised techniques for testing them. Dogs and fish, for instance, rely on chemical rather than visual cues to recognize themselves. In time, the gap might vanish entirely, or at least become considerably narrower.

For those species that do pass the MSR test, mirrors can be used to explore other aspects of their minds, as Reiss planned to do with her new experiment.

After watching the dolphins' training session, she and I headed down a flight of stairs to the base of one of the three large pools to get a closer view of the dolphins as they swam underwater. While we walked, she explained that she hoped to discover if self-awareness follows a similar developmental path in humans and dolphins. To do that, she needed to find out at what age dolphins acquire the ability to look in a mirror and say, essentially, "Hey, that's me."

Human children recognize themselves in mirrors when they're between eighteen and twenty-four months of age. At about the same age, they begin to develop feelings of empathy and become concerned about other people's feelings. They also become sensitive about having people look at them. They start to use personal pronouns and to play imaginative pretend games.

"All these developments are linked," Reiss said. "If you recognize yourself in a mirror, you understand you are an individual who is separate from others."

And that, in turn, psychologists say, gives you the ability to look at things from another person's perspective and to recognize that they, too, are separate beings, with their own thoughts and emotions. It is because of our extreme self-awareness that we reach out to help when we hear about people harmed by disasters; it is why, according to psychologists, we have feelings of sympathy and empathy.

Since dolphins have passed the mirror test, can we conclude that they are also sympathetic and empathetic, capable of behaving in similarly kind and understanding ways?

Reiss thinks so. "It could explain why dolphins help each other, and possibly rescue people from drowning and shark attacks."*

---

* Dolphins in the wild have also killed people, which is understandable in cases where people have tormented them, poking sticks in their blowholes, as one Brazilian man did

Many of these accounts are similar to one that four swimmers reported to a newspaper in New Zealand a few years ago. The group was swimming together off the coast of New Zealand's North Island when a pod of dolphins began to circle them, herding them together. One swimmer, Rob Howes, attempted to break free, but two of the largest dolphins pushed him back—just as he spotted a great white shark steaming his way. The dolphins, Howes said, had intentionally encircled them, protecting the swimmers before they even knew they were in danger. It could be, as some scientists suggest, that dolphins interfere with shark attacks because sharks are their mortal enemy and not because they're trying to protect people. Or, perhaps, swimming humans look like a struggling calf, and the dolphins, making an honest mistake, swim over to help out. This might explain why dolphins came to the rescue of Elian Gonzalez, a six-year-old Cuban boy who survived after the boat he and his mother were traveling on capsized and sank. His mother's boyfriend had placed him in an inner tube, he told his rescuers, and whenever he began to sink below the water's surface, dolphins would push him back up.

Researchers have never been on hand to observe what exactly the dolphins are doing in these situations, and they can only speculate about the reasons for the dolphins' behavior. But they've witnessed dolphins helping one another on numerous occasions. One of the first scientific reports recounts an incident that occurred on October 30, 1954. A team from a public aquarium was attempting to capture a bottlenose dolphin off the coast of Florida. They set off a stick of dynamite underwater near a pod of dolphins. One of the dolphins was badly stunned and having trouble surfacing. Immediately, two other dolphins came to his aid. They placed their heads beneath the injured dolphin's pectoral fins, keeping him buoyed at the surface "in an apparent effort to allow [him] to breathe," the scientists wrote. Meanwhile, the other members of the pod lingered nearby. In spite of the explosion, they did not swim away. They waited until their companion recovered—only then did they race off, leaping over the waves to speed their escape. "There is no

---

recently. The dolphin hit the man with such force that he died. What we don't expect is to be rescued by another species.

doubt in our minds," the researchers concluded, "that the cooperative assistance displayed for their species was real and deliberate."

There are also well-documented cases of other cetaceans coming to the rescue of unrelated species—which is another reason to think that dolphins' helping people is not impossible, although the question remains if they are doing so intentionally. In 2009, biologists watched a humpback whale in the Antarctic save a seal from the killer whales that were about to devour it. The orcas had washed the seal from an ice floe where it was resting, and the seal was leaping through the water, trying desperately to escape. Just then, a humpback surfaced nearby. She thrashed her tail flukes at the orcas, then rolled on her back and swept the seal onto the sanctuary of her broad belly. "Save the seal!" wrote the scientists who witnessed this remarkable behavior. Observers once also watched smaller cetaceans, probably pilot whales, rush to rescue a gray whale calf from killer whales off the coast of California. "We saw a large gray whale with her smaller youngster, and the youngster was being attacked by three orcas," wrote one of the eyewitnesses. Suddenly, more than a dozen pilot whales (which are about half the size of an orca) rushed in. They picked up speed to slam into the killer whales' heads, hitting them around the eyes over and over again until, after a good twenty minutes of incessant pounding, the killer whales swam off—and the gray whale mother and calf escaped.

Reiss is not surprised by these stories. "I've seen it," she said. "Two dolphins will lift a third to the surface to breathe, and sometimes if a calf dies, the mother will often stay with its carcass for hours."

We stopped beside the pool's window and peered inside. The water was a pale greenish blue, and in the distance we could make out the sleek, gray shapes of two adult dolphins and their calves.

"There are only mothers and young calves in this pool," Reiss said. "It's like a dolphin kindergarten."

In the wild, she explained, dolphin mothers and calves often spend time together in groups much like these. Although it's not known if dolphins intentionally join these groups, by doing so the calves have playmates, and the mothers have babysitters.

Just then, one of the mother-calf pairs swam up to the window to peer out at us. They pressed their eyes close to the glass, as if trying to get a good look. I had never seen a dolphin or a dolphin's eye that close and wondered at the intensity with which they studied us. Their eyes were round and as large as silver dollars, and their irises were milk-chocolate brown. The mother was pointed downward, like a sleek missile, occasionally sculling to keep close to the glass, while her calf tilted upward in the opposite direction.

Only the glass wall separated us, but it seemed we and the dolphins were reading each other's eyes and faces, looking for clues that might reveal something about our thoughts and intentions—a kind of eyeball-to-eyeball dialogue.

"They want to say hello," Reiss said. "Hello, Nani! Hello, Beau!" she called out, waving a hand at the dolphins. Nani, she added, was the mom, and Beau was her three-month-old son.

Self-awareness presumably confers the ability to do things intentionally—just as Nani and Beau were doing. They had swum over specifically to say hello to us, "to check us out," as Reiss put it.

We smiled at the dolphins, and of course I had a warm feeling toward them and felt a little surge of pride because they had chosen us.

"Hi, Nani. Hi, Beau," I said, echoing Reiss. I beamed at the dolphins as you do when making new friends.

Then Nani turned away, taking Beau with her. She swam to the surface and slapped the water hard with her tail, creating a small tsunami. The wave shot up and over the edge of the tank, and with stunning precision crashed down on our heads. It was as if someone had dumped a bucket of water on us. We were drenched to the bone.

"Oh!" we cried, as cold water spilled down our necks and pooled at our feet.

"I should have known better," Reiss said, pursing her lips and wringing her hair.

I fanned my notebook to dry the dripping pages, peeled off my wet jacket, and used a dry corner of my sweater to blot the inky, running notes. So much for empathy, I thought crankily.

In the pool, Nani sailed by with Beau in tow, churning the water and seeming to taunt us with her smooth, muscular moves. I didn't know if her perfectly aimed splash had taught me anything about reading a dolphin's mind and intentions—but this time I was sure there was a smirk, even a look of glee in her eye.

AFTER WE DRIED OFF, Reiss set up her experiment in a small room that was wedged like a mine shaft between the three dolphin pools. We climbed down a short ladder to reach the room, which had three narrow, rectangular windows. Each window looked into one of the pools. Standing here, you could watch all the dolphins in their tanks—and not get splashed. Reiss mounted a lightweight two-way mirror made of acrylic in one of the windows and blocked the other two with cardboard to make the room as dark as possible.

Sue Hunter, the aquarium's director of animal programs and a trainer with over twenty years' experience with dolphins, joined us as Reiss readied her video camera, focusing it on the back of the mirror.

"Okay, I'm going to start filming, so we'll have to be quiet," Reiss said. "Only whispers."

Through the mirror, we could see into the dolphins' tank, where five dolphins—three adults and two youngsters—were swimming. (Nani and Beau weren't in this pool.) It was the first time any of these dolphins had seen a mirror, and at first they gave it a wide berth.

"That's Chesapeake and her three-month-old calf," Hunter said softly, pointing to the two as they swam past the mirror, the calf nestled against her mother's side. "We just call her 'C.C.' for 'Chesapeake's Calf.'"

Behind that mother-calf pair came Jade and her year-old son, Foster. Shiloh, the group's dominant female and Chesapeake's mother, zoomed along in their wake. Foster was by far the liveliest and most energetic of these dolphins, "our upcoming superstar," Hunter said, and each time he rocketed by the mirror he inched closer to it. The others, particularly Chesapeake and C.C., kept their distance.

"They're doing flybys," Reiss whispered. "They're trying to figure out if those are other dolphins they see in the mirror."

In their social interactions, dolphins regularly mimic and match the moves of their friends, diving and leaping together like synchronized acrobats. The first time dolphins encounter a mirror or other reflective surface, they respond as if they were seeing other, unfamiliar dolphins, and they are cautious. In the wild, strange dolphins can be dangerous, particularly for a mother with a calf. There are a few accounts of dolphin males committing infanticide, killing the calves of females they don't know and have not mated. After losing her baby, the mother soon comes into estrus, giving the killers a chance to father her next calf. As horrific as it is to us, infanticide is a reproductive strategy found in many mammalian (and some avian) species, primarily those such as dolphins, where a female has a baby only every four to five years.*

Chesapeake needed to be wary as she and her calf swam past their mirrored reflections. Was she looking at dolphin friends or foes?

Many animals initially react to their reflections in a similar manner. You may have seen a male robin or other male songbird repeatedly attack his reflection in a window. Usually this happens during the breeding season, when males want to keep rivals out of their territories—and away from their mates. A male songbird spotting his reflection thinks that he sees an invader—and a very bold one at that. No matter how many times the defending robin rises up to furiously flap and peck at his intruder, the rude fellow always returns. It's enough to give a robin nightmares, especially since he never seems to grasp that he is attacking his own image. Some people might regard this as a sign of stupidity on the robin's part, but he has only a few short seasons to mate and pass on his genes. A smart male robin, who wants to make sure the eggs in the nest are his and not those of some passing Cassanova, can't waste time debating, "Is this guy for real?"

Other animals react similarly when they see their reflected images. As a youngster, our collie would growl at his reflection when the doggie-in-the-

* Female dolphins mate with numerous males, so the males don't know if a calf is theirs or not. All they can know is that they mated with certain females. As far as scientists know, male dolphins do not harm the calves of the females they've had sexual relations with.

window wouldn't go away. But ultimately, it seems dogs grow bored with their unidimensional, unresponsive adversaries and ignore them. (There is anecdotal evidence that dogs can use mirrors to find food, an ability called mirror-guided behavior.)

Dolphins behave differently after they realize their reflections are not dangerous, Reiss explained. They begin to slow down when they approach the mirror, as Jade and Foster were now doing, and to treat their images in a social manner, as if they were viewing another dolphin. They bobbed their heads in greeting. Of course, their reflections nodded back, mirroring their every move—which to dolphins is a way of saying, "I want to be your friend."

"That's the next thing they have to learn: that their reflection is really *not* their new very best friend," Reiss said, "so they do more exploratory behaviors," pushing the mirror and even echolocating on it.

In his first exploratory move, Foster swam in close and bumped our window mirror, then rubbed his body against it, causing a loud squeak. He shot off at the sound and made an extrafast lap around the pool. Then all five swam by in a tight pack. Shiloh returned alone and stopped in front of the mirror. She swam toward it, then backed up and nodded her head up and down.

"That's contingency testing," Reiss whispered. In other words, Shilo was experimenting with what the mirror (or dolphin in it) could do. Jade watched from one side, trying to sneak a peek, but like a competitive sister Shilo pushed her away. Shilo wanted the mirror to herself and hovered in front of it for several minutes, as if admiring her reflection. When she finally swam away, Jade at once took her place. She sidled up to the mirror, opened her mouth, closed it, and opened it again.

"You can really feel their brains at work," Reiss said. "She's figuring it out—that maybe it isn't another dolphin."

Just as humans do when they encounter a mirror for the first time, the dolphins now began to use the mirror to regard and study themselves. Foster swam close to the mirror, stopped and slightly opened his mouth, then stretched his jaws apart a bit farther, while turning his head to get a better view. Finally, he opened his mouth as wide as possible, showing his white, peg-shaped teeth and fat, pink tongue. We three humans covered our mouths with our hands to stifle our laughter. Isn't this what mirrors are made for: to peer into those critical places you would otherwise never see?

There were other body parts to explore, and Foster turned upside down to look at his blowhole. He twisted to the left and right and blew out a stream of bubbles. Once more, he stood in front of the mirror, bobbing his head and opening his mouth like a kid ogling himself in front of the funny mirrors at a carnival.

For twenty minutes, we watched as the dolphins experimented with the mirror. Only Chesapeake and C.C. did not venture too close to it, although on some of their laps Chesapeake didn't stop her calf from slowing down and giving it a long look.

When the videotape ran out, Reiss stood up and stretched. "That was great," she said, pleased with this first step in her new experiment. "They all reacted much more strongly than I would have guessed. And I'm surprised at how fast Foster started stationing himself in front of the mirror" to investigate his reflection.

She would return in a few weeks to let the dolphins explore the mirror again and, sometime after that, to give the mark test to Foster and C.C.

Did Foster's behavior suggest that he knew he was looking at himself?

Reiss hesitated. "I wouldn't be comfortable saying that, not until I give the actual test. But he is very relaxed with the mirror, and I didn't expect that to happen so quickly."

If Foster did recognize himself at this young age, it might mean that dolphins develop this sense of self early in their lives—which "makes sense given how social and gregarious dolphins are," Reiss said.

Reiss, who is normally very chatty, fell silent as she packed her gear. I thought she was worried about the train she had to catch and, to help out, offered to call a taxi. She shook her head. "It's like entering another world when I'm watching the dolphins," she said, half apologetically. "I get lost in it, and it takes me a while to come back out."

I had to agree: there was something mesmerizing, almost seductive, about watching the dolphins, and after Reiss left I lingered near the dolphins' pools with Hunter. It was more than just the ease with which the dolphins swooped through the water, or how they listened so attentively to their trainers. Something about them seemed uncannily savvy. I wondered if, after her many years of training dolphins, Hunter knew what I meant.

"They're thinking," Hunter said simply. "You can see it especially when they're learning something new. They're different from us, and I'm not a scientist, so I wouldn't venture to say how they're thinking. But they're thinking."

Psychologists have recently shown that we humans are adept at merging our minds with one another. People having a conversation experience "neural coupling," meaning that similar parts of our brains light up as we speak and listen to one another—a phenomenon they discovered by using magnetic resonance imaging to compare the brain activity patterns of speakers and listeners. We may also engage in something like neural coupling when watching actors or reading novels—those moments when we feel as if we're experiencing the actions and emotions of our favorite characters. Some researchers suspect that neural coupling occurs because these activities engage our mirror neurons—cells that fire when we watch or imitate the actions of another.

Can neural coupling also explain what happens when we watch certain other species, especially those with specialized brain cells that are similar to ours? The brains of humans, elephants, and dolphins and other cetaceans have specialized spindle cells that are thought to be important in complex social behaviors. Scientists also suspect that dolphin brains have mirror neuron cells because these cetaceans are such accomplished mimics. When we gaze at dolphins and they gaze back at us, are we engaging in some type of cross-species neural coupling?

Whatever the reason, people around the world in numerous cultures have felt a kinship with dolphins for thousands of years. We swim with them, summon them in some cultures to help us catch fish, weave them into our myths, and (in modern times) revere them as wise therapists. But only since the late 1950s have we tried to understand their minds scientifically; scientists have studied the brains of rats for more than twice as long. What do we really know about how dolphins think? As it happens, quite a bit, as I learned in Hawaii, where I joined Louis Herman, one of the pioneers of dolphin cognition studies and the director of the Dolphin Institute—the "home of the world's most well-educated dolphins."

IT WAS RAINING when I ducked into the Dolphin Institute to talk with Herman about his dolphins.* The institute was housed in temporary quarters—a single, beige metal building—on the edge of a grassy field at Ko Olina, a large resort on the west coast of Oahu Island. The ocean lay far in the distance, looking dark and dull on this stormy day. There wasn't a dolphin in sight.

Herman met me with a warm handshake. Fit and tanned from his daily ocean swims, he is of medium height, with silvery gray hair and blue eyes. He has a New Yorker's accent and a New Yorker's intense manner, too, despite living almost forty years in Hawaii. It was easy to picture him as the taut, impetuous, and hot-tempered young man who had once slugged a colleague at a marine mammal conference for taunting him about the loss of two of his dolphins.

Herman was eager to talk about dolphins and dolphin brains, and launched into a rapid-fire explanation about their special qualities as soon as I walked in the door.

"Dolphin brains fascinate people, and you can see why," Herman said, grabbing a plastic model of the brain of a dolphin from a shelf. "Just look at all the fissures and folding."

The model had the shape of a slightly flattened melon and the texture of a large piece of coral. I ran my fingers over the undulating cracks and crevices, wondering what kind of mind a real one would hold. Dolphins, after all, are creatures of the sea, and they use sonar for navigating, a sense we do not possess. Yet they are mammals, too, and as with all animals, including us, their brains are designed primarily for solving life's basic challenges: eat, avoid predators, find a mate, reproduce. So why have dolphins and humans evolved such large brains, which are energetically expensive to grow and maintain?

* The institute was originally called the Kewalo Basin Marine Mammal Laboratory and was affiliated with the University of Hawaii; the lab no longer exists.

"We'll talk about that," Herman said. "But just look at all these folds and convolutions; they give the dolphin's cortex a surface area that's larger than our own. It's not as thick as ours. But a dolphin's cerebellum is much larger. And their brains are twenty-five percent heavier than ours, too."

The brains of bottlenose dolphins score high on several measures, Herman continued. It's not just that their brains are large, which is expected in an animal of their size. Dolphins have one of the highest ratios of brain size to body mass in the animal kingdom. Measured in this way, a dolphin's brain is second in size only to that of a human's, and well above that of our closest genetic relative, the chimpanzee.

A large brain-to-body size ratio is often considered a hallmark of intelligence, although it's not the only factor. But it is something we humans celebrate in ourselves and regard as key to the flowering of human intelligence. So it's easy to imagine scientists' eyebrows arching in surprise when dolphins—and not chimpanzees—were discovered in the 1950s and 1960s to have a brain nearly as large as our own on this scale. Although humans and dolphins are both mammals, we are as distantly related as elephants and mice, and we haven't shared a common ancestor in nearly one hundred million years. They evolved their large brains millions of years before our ancestors did—which means these seagoing mammals already had experienced an explosive surge in brain capacity when our thick-browed relatives were still climbing trees.

Why did dolphins evolve such large brains? What are they doing with all that gray matter? And was it because of this intriguing similarity between dolphin brains and those of humans that Herman chose to study these creatures?

"No," Herman said, ushering me into a nearby conference room. "I came to study them purely by chance."

We each took a seat at a long table, where he had arranged a stack of his research papers and a projector so that we could watch films of dolphins during various cognition tests. But before we watched these I stopped him. I was just a bit puzzled, I explained, because I had seen photos on the institute's website of Herman's dolphins swimming in two Marineland-type tanks. Where were these?

Herman shook his head. "This isn't where we did that research," he said. "We aren't working there now." He seemed reluctant to say more, as if the subject was painful. He was silent for a moment and then picked up his story about how he first began to work with dolphins.

"I'm not a marine biologist," Herman said. "I'm a human cognition psychologist. That's what my training is in, studying how humans process information." His dolphin research was what he called his "unintended career," because prior to working with them his main experience with animals had been studying rhesus monkeys when he was a graduate student at Emory University.

Herman's position as a human psychology professor at the University of Hawaii was unintended as well. A few years after completing his doctorate at Penn State in 1961, Herman was thriving in his career as an assistant professor at Queens College in New York City when he fell afoul of the college's internal politics. Desperate for a job, he spotted an advertisement for the position in Hawaii and was hired. He and his wife, Hannah, also a psychologist, moved to the island in 1963. "We thought we'd be here a couple of years at most," said Herman, who after arriving in Hawaii continued to pursue his research on human cognition. Then, because he had experience working with animals—the monkeys at Emory—the department asked him to teach an undergraduate lab course on the principles of learning, using white rats.

One semester, at the end of every lecture, a serious-eyed young man accosted him in the hallway. "He always confronted me with the same question: 'Why, in Hawaii, are you studying rats?' He thought I should be studying dolphins. The idea had never occurred to me."

In those days, most cognition researchers held dolphin studies—and dolphin scientists—in low regard because of the work of John C. Lilly, a neurophysiologist. In the 1950s, Lilly helped map the cortex of the bottlenose dolphin, and later he became captivated by their vocalizations. He was convinced that their clicks, buzzes, and whistles constituted a language and that his study dolphins were even trying to communicate with humans in "delphinese."* He also undermined his credibility by giving LSD to the dol-

* Lilly was one of the first to describe dolphin vocalizations, and his 1961 paper in *Science* about their sounds is still considered valid, although his later idea about "delphinese" is not.

phins to see how this affected their behavior and by claiming that dolphins had oral histories extending back twenty million years. Lilly's ideas and writing style were engaging, and his books were (and are) popular with the general public, but not with rigorous scientists, who blame him for creating a mythical and mystical image of dolphins that persists to this day.

Herman was well acquainted with Lilly's ideas, as was the student who wanted Herman to give up rats for dolphins. "He always mentioned John Lilly to me," Herman said. "So my first inclination was to say, 'Yes, how interesting,' and keep walking. But as to why I was studying rats in Hawaii—I really didn't have a good answer for that."

Eventually, Herman yielded to his student's constant prodding and looked into what other scientists knew about dolphins. The field was in its infancy, but some researchers, particularly David and Melba Caldwell, were doing solid work. The Caldwells had discovered that each dolphin has a unique acoustic signature—in other words, a whistle call that functions like a name, exactly like the signature contact calls of the green-rumped parrotlets. Other scientists, also working with captive dolphins, revealed that dolphins hunted via echolocation. All the scientists studying dolphins seemed enchanted by how sociable, gregarious, and playful dolphins were in captivity; trainers loved working with them because, as Sue Hunter had told me in Baltimore at the National Aquarium, dolphins are good students, quick to learn.

Deciding that dolphins might be a good species after all to use in a study, Herman challenged his graduate students in the summer of 1967 to collectively design an experiment on dolphin intelligence. The university did not have dolphins for research, so Herman turned to Karen and Tap Pryor, who had opened Sea Life Park, a commercial venture, on the island a few years before.* The pair eventually agreed to let Herman and his students use one female dolphin, named Wela, for their project. "We wanted to see if dolphins could surpass the learning and problem-solving abilities of rhesus monkeys," Herman said. By the summer's end, Wela was as good as a macaque at solving

* Karen Pryor is a self-taught animal trainer and was the curator of Sea Life Park's mammals. She later earned fame as an animal behavior scientist and as a writer and promoter of the clicker method of training that is widely used by dog, cat, dolphin, and many other trainers.

the scientists' tests, and Herman suspected that she was probably better—but because the Pryors had agreed to let him use Wela only for the summer, Herman was running out of time to show just how good she was.

Then serendipity struck. On one of his last research days, a small group from the U.S. Navy's newly launched marine mammal program stopped by to watch Herman work with Wela. Herman explained the details of his students' study, adding that it was soon to end. The leader of the naval group, Bill Powell, asked if Herman or the university had a place to continue the research.* Herman shook his head. Powell was silent for a few moments and then said that if Herman could find such a place, Powell would provide him with two dolphins and some funding.

"That changed my life," Herman said—and the field of dolphin research.

Herman searched for two months before finding an abandoned shark exhibition facility at Kewalo Basin Harbor, which he renovated for dolphins. A few months later, on March 4, 1969, a navy truck delivered two Atlantic bottlenose dolphins, Keakiko (Kea, for short) and Nana, to Herman's care, and the project was under way. Nana was suffering from an untreatable pox virus, and to Herman's sorrow she died a few months later. Three years later, he acquired another dolphin, Puka.

Herman would never again study human cognition or use rats for lab classes. For the next thirty-three years, he would explore what he'd glimpsed in Wela—the quicksilver intellect of the bottlenose dolphin.

"Lilly thought dolphins have large brains in order to communicate," Herman said. "But they also have large brains for the same reason we do: they are long-lived and have socially demanding societies. Like a human child, a young dolphin has a great deal to learn."

Herman chose not to focus on what dolphins need to learn in their natural societies. Instead, like a parent of an especially bright child, he wanted to explore what a dolphin is mentally capable of achieving, and he developed what he called his "guiding philosophy."

"We wanted to bring out the full flower of the dolphin's intellect, just as

---

* Bill Powell also went on to study dolphins, becoming in 1968 the manager of the U.S. Navy's dolphin program at the Marine Corps Air Station at Kaneohe Bay.

educators try to bring out the full potential of a human child," Herman said. "My thought was, 'Okay, so you have a pretty brain. Let's see what you can do with it.'"

In time, Herman acquired four other young Atlantic bottlenose dolphins, Akeakamai, Phoenix, Elele, and Hiapo. They lived in the two large, circular outdoor tanks at the facility he had renovated and renamed the Kewalo Basin Marine Mammal Laboratory. That's where the photographs on the website had been taken, Herman explained.

It didn't take long for the dolphins to captivate Herman. Curious and playful, they transferred their sociability to Herman and his students. Herman didn't attempt to decipher the dolphins' whistles and clicks but instead devised a hand-and-arm gestural language to communicate with the dolphins. His invented language had a simple grammar, so that Herman could construct sentences and test if the dolphins actually understood what he was saying to them.

For instance, a pumping motion with his fists closed indicated the word "hoop"; extending both arms overhead (as in jumping jacks) meant "ball." A "come here" gesture with a single arm told the dolphins to "fetch." Responding to the request "hoop, ball, fetch," Akeakamai would push the ball to the hoop. But if the word order was changed to "ball, hoop, fetch," she would carry the hoop to the ball. Over time, she could interpret more grammatically complex requests, such as "right, basket, left, Frisbee, in," asking that she put the Frisbee on her left in the basket on her right. If he reversed "left" and "right" in the instruction, Akeakamai would reverse her actions.

Herman switched on his film projector and narrated as we watched Akeakamai responding to his sign-language instructions.

"Once she understood this language, she could respond to new requests the very first time we made them," he said. "These weren't trained behaviors. She had a deep understanding of the grammar of the language."

Dolphins are also a very vocal species (they are vocal learners), and they could easily imitate the random sounds—whistles or electronic, pulsing noises, for instance—that Herman broadcast into their tank. "That's probably tied to their own need to communicate," Herman said. "I'm not saying they have a dolphin language. But they are capable of understanding the novel instructions that we convey to them in a tutored language; their brains have that ability."

There were other things the dolphins could do that "people have always doubted about animals," Herman said, as the dolphins in the film dived to look at a television screen behind an underwater window. On the TV's screen, a person gave them gestured instructions.

"They correctly interpreted those instructions, and on the very first occasion," Herman explained. "They recognized that television images were representations of the real world that could be acted on in the same way as in the real world. And they'd follow a trainer's instructions on the TV."

They readily imitated motor behaviors of their instructors, too. If a trainer bent backward and lifted a leg, the dolphin would turn on its back and lift its tail in the air, creating an analogy between its tail and the human leg. Although such physical imitation was once regarded as a simpleminded skill, in recent years cognitive scientists have revealed that it is extremely difficult, requiring the imitator to form a mental image of the other person's body and pose, then adjust his own body parts into the same position—actions that are thought to require mirror neuron cells. And, like recognizing one's image in a mirror, the ability to imitate implies an awareness of one's self, psychologists say.

"Here's Elele," Herman said, showing a film of her following a trainer's directions: "Surfboard, dorsal fin, touch."

Without hesitating, Elele swam to the surfboard and, leaning to one side, gently laid her dorsal fin on it, an untrained behavior. The trainer stretched her arms straight up, signaling "Hooray!" and Elele leaped into the air, squeaking and clicking with delight. To accomplish this task, Elele had to understand the gestured directions and use a body part—her fin—that she could not see but pictured in her brain.

"Elele just loved to be right," Herman said. "And she loved inventing things. We made up a sign for 'create,' which asked a dolphin to create its own behavior."

Dolphins often synchronize their movements in the wild, such as leaping and diving side by side, but scientists don't know what signals they use to plan their moves or to stay so tightly coordinated. Herman thought he might be able to find out how with his pupils. In the film, Akeakamai and Phoenix are asked to create a behavior and do it together. The two dolphins swim away from the side of the pool, circle together underwater for about ten seconds, then leap out of the water, spinning clockwise while erect and squirting water from their mouths, every maneuver done at the same instant.

"None of this was trained," Herman said, "and it looks to us absolutely mysterious. We don't know how they do it—or did it."

He never will. Akeakamai and Phoenix and the two others died in 2003 after developing an infection that did not respond to treatments. This was why there were no dolphins at the Dolphin Institute.

"I might try again," Herman said. "It's why we're here in these temporary quarters. We're thinking about a new program. But . . ." He shrugged, and his voice trailed off. He was an emeritus professor now, and busy with another study of humpback whales in the wild. He wasn't sure about beginning again with captive dolphins. And he had gathered so much material from his dolphin study that he still was writing papers about it.

"I loved our dolphins," Herman said, leaning forward, "as I'm sure you love your pets. But it was more than that, more than the love you have for a pet. The dolphins were our colleagues. That's the only word that fits. They were our partners in this research, guiding us into all the capabilities of their minds. When they died, it was like losing our children."

Herman pulled a photograph from one of the files. In it, he is in the pool with Phoenix, who rests her head on his shoulder. He is smiling and reaching back to embrace her. She is sleek and silvery with appealingly large eyes, and she looks to be smiling, too, as dolphins always do. It's an image of love between two beings, two vastly different creatures—one of land and one of sea—having what Herman said he had most wanted with his dolphins: "a meeting of the minds."

Herman studied the photograph and slipped it back into the folder.

We walked outside and drove to a restaurant overlooking the ocean for a late lunch. There Herman told me about the fate of the two other dolphins, Kea and Puka, he'd first worked with. They were as bright and engaging as the other four, and he had been on the verge of a major discovery about dolphins' comprehension of language, when two of his former students, who were in their late twenties and had worked for him part time as tank cleaners, decided to set the dolphins free. Herman suspects they did so because he had "discharged" them from their cleaning jobs. Two days after they were fired, the pair returned to the lab—but in the dark hours of the morning. They loaded Kea and Puka into a VW van, drove them fifty miles to a remote location, and released them into "waters unknown to them," Herman said.

"These were Atlantic bottlenose dolphins, they knew nothing about the Pacific Ocean, and they had been in captivity all their adult lives. I can't imagine what it was like for them to be dumped into these waters—lost and alone."

Herman arrived at the lab after a frantic student telephoned to tell him about the abduction. A throng of news reporters was already there, but at first no one knew where the dolphins might be. Herman doesn't remember who, but someone at the lab or among the reporters heard that there was a dolphin near shore and hanging out with swimmers at a site fifty miles away called Makua. "Somehow I knew it must be one of mine," Herman said. As he raced to his car, one of the weather and traffic reporters offered to fly him in his helicopter to Makua.

Herman spotted his dolphin, Kea, from the air. (Some snorkelers say they saw Puka several times over the next few weeks, a notion that Herman scoffs at.) The reporter hovered over the rock jetty, and Herman clambered down. "I ran to the beach and into the water, stripping down to my bathing suit, and swam out to Kea. She was several hundred yards out, apparently seeking some comfort from being near to the swimmers. She clearly recognized me, and I was able to give her two or three fish before she refused any more. That's a typical dolphin response to stress—to stop eating. She would not let me, or anyone, touch her. She had been manhandled, stressed, even traumatized by her experience, and now seemed afraid of everyone. So I just swam with her in the ocean, trying to keep her calm until help arrived."

Eight years before, when Kea first arrived at Herman's lab, he had immediately noticed a large shark scar that ran down her back to her underbelly; she had also been missing part of one flipper that the shark had bitten away. "I was amazed that she'd survived that," he said, "and when I found her in the ocean, I thought about that attack. I wondered whether she remembered it. I thought how frightened she must be. She was circling aimlessly and whistling—making distress whistles—continuously." One of Kea's eyes was badly injured and swollen shut, the result of the long, bouncing ride on the floor of the van after her abduction, and she was bloodied from washing against the rocks at the jetty.

Herman thought that help was on the way. The reporter had flown back to the lab to tell Herman's assistants, but they misunderstood the news; they thought Herman already had Kea safely on the beach. Instead of bringing nets to rescue her, they brought only a dolphin stretcher. "Then there was nightfall, a long, dark wait on the beach until dawn, and then a fruitless attempt the next day to bring her to shore—and then she was gone.

"I never saw her or Puka again."

We sat silently, looking out at the Pacific and its dark waves. They rose and fell—all the way to the horizon.

8

THE WILD MINDS OF DOLPHINS

If an alien came down any time prior to about 1.5 million years ago to communicate with the "brainiest" animals on earth, they would have tripped over our own ancestors and headed straight for the oceans to converse with the dolphins.

LORI MARINO

Much of what we know about the nuts and bolts of dolphins' mental abilities comes from captive studies like Reiss and Herman's simply because it is easier for scientists to work with these large marine mammals in a confined setting. It is through captive studies that we know, for instance, that dolphins possess excellent short- and long-term memories; that their eyesight is as good as ours both in and out of the water; and that their echolocation allows them to experience their surroundings more acutely than we can—they can "see" their world acoustically. A trained dolphin can use its biosonar to detect a steel ball just an inch in diameter that's nearly a football field away. Captive dolphins have also given scientists some glimpses of their playful, creative, and inventive sides; some have even been clever enough to outwit their trainers. A dolphin named Kelly at the Marine Life Oceanarium in Gulfport, Mississippi, was taught to pick up any litter that landed in her tank and give it to her trainer in exchange for a fish. Soon Kelly began surfacing with more and more bits of trash. Puzzled, the trainer asked a pool engineer to investigate. The engineer discovered that Kelly had accumulated her own private stash of trash. She kept it stored beneath a rock and retrieved pieces from her pile whenever she wanted a fishy reward.

Scientists also got their first hint about the larger reason for dolphins' brainy behaviors by watching those in captivity. Arthur McBride, the pioneering dolphin curator at Marine Studios in Florida (which later became Marineland), first suggested in 1948 that dolphins' cognitive talents must be driven by their sociability. Once McBride's group captured two male dolphins for the Studios. The dolphins were together when they were seized but were separated for several weeks afterward. When the smaller dolphin was finally released into the larger male's tank, the larger male "exhibited the greatest amount of excitement," McBride wrote. "No doubt could exist that the two recognized each other." Other early dolphin researchers also commented on the strong social bonds among their captive dolphins and the amount of time they spent nuzzling, stroking each other with their fins, and just hanging out with one another. It was all so . . . humanlike. But how was their sociability related to their intelligence?

McBride couldn't answer that question by observing captive dolphins. Just as it's not possible to understand why parrots evolved vocal mimicry by studying them in cages, watching dolphins in a tank won't shed much light on how or why they became so sociable and bright in the first place. It will tell you nothing about the ecological and social pressures that shaped their evolution and produced such an intelligent "mind in the waters"—the poetic phrase that the writer and conservationist Joan McIntyre first used in 1974 to describe the cetacean brain.

To understand the social nature of dolphins—and how being social can lead to intelligence—you need to spend time with dolphins in the wild, studying them just as Jane Goodall has studied wild chimpanzees.

For twenty-five years, Richard Connor, a cetacean biologist at the University of Massachusetts, has been following the antics and behaviors of wild bottlenose dolphins at his research site in Shark Bay, Western Australia. His is not the longest-running study of wild dolphins. In Florida, Randall Wells has been observing communities of dolphins that live in Sarasota Bay since 1970. Connor's site has one big advantage: the waters of Shark Bay are crystal clear, making it possible to witness dolphin behaviors from a boat. Connor and his students don't swim with their dolphins because that would interfere with the dolphins' normal routines and because, as the bay's name suggests, sharks are plentiful. Nor have they carried out experiments with the dolphins. Their study of the dolphin mind is largely observational—a type of research that requires long years of watching and recording an animal's behaviors, analyzing these statistically, and ultimately drawing conclusions about what forces drive their decisions and actions. It is classically old-fashioned natural history, still the best method for discovering why an animal species needs intelligence to survive and reproduce. Connor, who's in his fifties and who has, as one colleague put it, "a mind the size of a planet," is a master at it.

Nearly every summer, Connor migrates from his grand Victorian house in Massachusetts to a trailer park at Monkey Mia, a resort and campground on the shores of Shark Bay. He settles into an old RV that sits next to a weather-beaten house trailer, where his student assistants live year round. In the summer of 2009, he had three such helpers, Anna Kopps, Kathrin Bacher, and Kim New. Kopps and Bacher came from Switzerland to join the project, and New made her way from Hawaii. They were all working on advanced degrees in marine biology; Connor's project gave them a chance to gain valuable field experience.

I met up with Connor and his team for an early breakfast in their trailer. We all squeezed around the small formica table to down bowls of cereal, yogurt, and fruit. Outside, the stars were vanishing into the dawn's lavender sky, and there wasn't the slightest breeze—perfect dolphin-viewing weather, Connor said. As long as there wasn't a wind to ruffle the sea's surface, the water would stay clear, and we could watch the dolphins as if peering through glass.

"That's what makes Shark Bay so remarkable for dolphin research," Connor said. "You'll see more behaviors here in a week than you can in a lifetime at other sites."

Connor is about six feet tall, with salt-and-pepper hair and the weathered complexion of a sea captain. He was dressed like an old sea hand, too, in a navy blue windbreaker, khaki-colored shorts, and flip-flops, his standard uniform for dolphin watching. His speech and avuncular manners, though, were those of a teacher, someone accustomed to guiding others and offering advice, not barking orders. He made it easy to fit in with his team; I could tag along like any other student, fascinated—as Connor has been for almost his entire life—with dolphins. He had missed the two previous seasons because his partner, an antiques collector, had fallen ill with a rare form of cancer and died.

"It was a rough time for me," Connor had said, when I first talked to him by telephone to inquire about the possibility of visiting his site. "I haven't been to Monkey Mia in two years, and I need to get back." He felt that he was on the cusp of discovering why dolphins have large brains and a cognitive repertoire to match. But he needed to be out with the dolphins, tracking who was with whom, to prove his hunch.

And now Connor had been at Shark Bay for three weeks. He smiled when I asked how it felt to be back.

"My first day on the boat was literally like an addict getting a fix," he said. "It was a huge release, incredible." He curled his fingers into the palms of his hands, making tight fists, as if to better grasp that moment. "I'd missed it all—the dolphins, the air, the water, the sun . . . this trailer." He stood up, abruptly. "Let's go find some dolphins."

Outside, the three women had already hooked up the boat trailer bearing their small, bright red motor launch, *Sponge Bob*, to the team's Land Cruiser. We jumped inside and, with Kopps at the wheel, drove the short mile to Monkey Mia's boat ramp. The team made this same trip every day, six days a week, if the weather held—yet there was an air of anticipation in the Land Cruiser, and lighthearted jokes about the day ahead and which dolphins we should visit, as if the Monkey Mia cetaceans were old friends waiting for us around a bend in the bay. We would be looking in particular for certain male dolphins and their pals, Connor explained, since they were the focus of his study.

Over the years, the researchers have identified and named more than a thousand dolphins in Shark Bay, using the different nicks, cuts, scars, and scrapes on their dorsal fins to tell them apart. The males travel together in groups, while the females are most often alone or with their latest calf; other researchers who collaborate with Connor keep track of the females' doings.

"The male dolphins here do have buddies," Connor said, as we pulled up to the boat ramp. "But they don't always have the same pal. They live in a world of shifting alliances, and that's what makes it mentally demanding.

They don't have big brains because they're searching for food. They have big brains for one of the same reasons we do: they're dealing with social uncertainty."

Connor's team monitors more than twenty male dolphin alliances in the bay, some made up of only two or three buddies and others that have a dozen or more partners. He has names for the dolphins and for their alliances. And as he called these out, it dawned on me that dolphins in the wild behaved far differently from what I thought. The dolphin alliance list included the Grand Pooh-bahs, and the Gas Gang; the Wow Crowd (named for one dolphin who had such an impressive scar on his back from a shark's bite that Connor had called out, "Wow!," when he first saw it) and Captain Hook's Crew. There were the Prima Donnas and the X-Fins; the Blues Brothers and the Kroker Spaniels; the Sharkies and the Rascals.

On this day, Connor decided we would motor north in Red Cliff Bay to look for the Prima Donnas, a seasoned alliance. We would follow them, record each dolphin's activities, and keep an eye out for other male dolphin alliances—and any wandering females, who might be in estrus. Connor's research had already revealed that dolphins are astute social thinkers, plotting whom best to ally themselves with. In some ways, he thinks they're cleverer social climbers than chimpanzees and may approach humans in terms of their tangled political intrigues and tactical dramas. During this season, Connor hoped to record more evidence of dolphin social scheming to support his idea.

As the researchers readied *Sponge Bob* for our outing, I tested the chilly waters with my bare feet and watched for dolphins. It didn't take long to spot one—in fact, there were several adults and two calves swimming in the shallows directly in front of the resort's restaurant and only a few feet from the sandy shore, where a throng of tourists had gathered to watch. Their dorsal fins shone like burnished pewter keels skimming above the sea's clear water, and their breaths exploded loud and wet—*whoosh!*—against the quiet of the dawn.

"They're some of the Beach Bums," Connor said, as I climbed aboard *Sponge Bob* and took a seat on the foredeck. "It's not an alliance; just some of the female dolphins who hang out here, waiting to be provisioned."

Monkey Mia is famous for the panhandling Beach Bums, who wait in just a few inches of water with their rostrums poking up in the air, while holding themselves upright with their pectoral fins—a posture guaranteed to get a human's attention and, for a lucky dolphin, a handout of mackerel. Some say that aborigines started feeding the dolphins long ago, while others attribute the dolphins' begging behavior to a softhearted, elderly woman who lived on a sailboat in the bay in the 1960s and fed the dolphins. However it developed, several dolphins turned into regular moochers. Later, as word spread about the friendly ways of the Monkey Mia dolphins, more people came to visit, and the feeding got out of hand. After some of the dolphins fell sick and died in the late 1980s, rangers from Australia's Department of Environment and Conservation stepped in. Now the dolphins get treats only two times each day, and only from people—usually kids—the rangers choose.

Old habits die hard, of course, and as Connor piloted *Sponge Bob* over the marina's waters, one of the Beach Bums swam right up to our boat.

"Well, hello there, Nicky—you old beggar," Connor called to her, as she turned on her side to gaze up at us. She lifted her beak slightly out of the water, showing her pearly teeth. Nicky was more than thirty years old (wild dolphins can live to about fifty) and she had been a Beach Bum for her entire life.

"Yeah. You'd like a fish, wouldn't you? But you're not getting one from us," Connor said.

Nicky followed us a short distance but soon turned back to the shore, where her prospects were better.

"Nicky is the daughter of Holeyfin, who was the most popular and well known of all the Beach Bums—but she's gone now," Connor said. "I met them both on my first trip here, way back in 1982."*

Connor had been twenty-four and an undergraduate at the University of California in Santa Cruz when he first arrived at Monkey Mia, along with a fellow student, Rachel Smolker. They'd heard about the wild, friendly dolphins from a visiting lecturer at the university and had subsequently sold all

---

* Holeyfin died in 1995 from a stingray's spine that pierced her heart. Her sleek, stuffed body is now on display in the Monkey Mia visitors' center.

their belongings to raise enough money to make the trip. Just after dawn on their very first day, as they watched the bay's waters, a dolphin with a calf surfaced a mere twenty yards offshore.

The young scientists didn't hesitate. They rolled up their pants and waded in, watching in amazement as the adult dolphin swam directly toward them. The dolphin, who had a small hole in her dorsal fin, lifted her head and turned it toward them, looking each of them in the eye. She opened her mouth slightly, as if grinning, and stopped beside their legs. She seemed friendly, and they found themselves reaching out—tentatively at first, then with growing confidence—to stroke her along her sides. After awhile, she turned abruptly and swam back to her calf.

"That dolphin was Holeyfin," Connor said, "and the calf was her two-year old daughter, Joy. But if you'd told me that I was going to be touching a wild dolphin on my first day here . . . Well, it was incredibly cool."

The owners of the Monkey Mia camp had named these two dolphins, and several others that were regular beach visitors, using various marks (like the small hole in Holeyfin's dorsal fin) to tell them apart. They also kept records of the dolphins' births and deaths, which they shared with Connor and Smolker.

"I didn't know very much about animal behavior then," Connor said, "but I could see that this could be the dolphin equivalent of Gombe [Jane Goodall's chimpanzee study site in Tanzania]. It was unreal: the easy access to wild dolphins, the clarity of the water."

Dreaming big, Connor and Smolker decided to give their venture an official name: the Monkey Mia Dolphin Research Project. (Connor now runs the Dolphin Alliance Project to support his research.) Only the project's name was grand. They did not have enough money to even rent a boat so that they could be out among the dolphins. Instead, that first season, they watched the dolphins almost entirely from shore, noting the dolphins' behaviors and relationships on handheld tape recorders. The recorders, an old hydrophone, for listening to the dolphins' underwater sounds, and two cameras (and black-and-white film, because they could not afford color) constituted all of their scientific equipment.

"We didn't have a clue about how to do scientific observations, but actually all you have to do is watch and pay attention," Connor said. "And by doing that, we learned how dolphins behave."

Connor and Smolker also learned about the dolphins' temperaments and personalities by feeding and petting them, something Connor never does today (the rangers also strictly forbid this). "It was really the way we learned to identify them as individuals," he said. Most of the dolphins were surprisingly good-natured and tolerant and enjoyed games like keep-away that the two researchers played with them using strands of sea grass. Some accepted more petting than others; none liked to be touched on their dorsal fins or blowholes or around their eyes.

"They had rules for how you should behave," Connor said, a discovery that would later inform his understanding of dolphin relationships. "They would warn you about something they didn't like. They'd snap their jaws or toss their heads. If you tried it again with some dolphins, like Nicky, you'd get bitten."

Toward the end of the summer, on two perfectly windless days, the camp owners let Connor and Smolker borrow what they most needed: a boat. For the first time, the two glimpsed the dolphin population in the deeper waters of Shark Bay, further opening their eyes to all that might be possible at Monkey Mia.

"We saw so many dolphins, it was phenomenal," Connor said, shaking his head at the memory. "We had no idea. And they were *so* tame. They didn't mind us at all. We met dolphins that day who, years later, would turn out to be extremely important to our study. We were out on the water from dawn till dusk, and we took photographs of every dorsal fin we saw."

Connor and Smolker were also able to identify the sex of many of these offshore dolphins because they had the happy habit of occasionally turning over to float belly up, exposing their genitalia—or at least, the slits that conceal these. Males have two slits—one housing the penis, the other the anus. Females have a single slit that encloses the vagina and anus; on either side of the genital slit are two smaller ones containing the mammary glands.

"That made a huge difference because right away we could figure out the males and the females," something that took months or longer to do at other dolphin study sites, where the dolphins aren't as accommodating and the water is not as clear.

Who were these offshore dolphins? Some adults seemed to be traveling in tight pairs or trios. Why was that? What were the relationships between these dolphins and the others that visited Monkey Mia? Connor and Smolker had neither the money nor the resources to answer their questions, but both vowed to return.

SMOLKER MADE IT BACK FIRST, arriving in 1984 with a small grant to begin a study of dolphin communication and social relationships. Connor returned two years after that with a grant of his own to explore the male dolphins' relationships for his doctoral thesis.

"The males' alliances are what you can most easily observe and quantify about dolphin behavior," Connor said. "They are the key to almost everything we've discovered here about dolphin cognition. And they're why we're out here today, updating our records. "

From Monkey Mia we motored northeast, running parallel to the red sandstone cliffs of Peron Peninsula. The peninsula divides Shark Bay into two U-shaped inlets. Connor's study area covers the northernmost inlet, about two hundred square miles. There were only a few other boats, mostly fishing vessels and small sailboats on the water. To the west the sky was streaked with ribbons of clouds, but ahead of us the sun shone brightly, if weakly. We all stayed bundled up against the morning chill and wind and watched the waters for dolphin fins. Connor was the first to spot them.

"Okay, who do we have up ahead?" Connor called to Bacher. He pointed in the distance, and as I followed his finger half a dozen fins came into view. They shimmered against the sunlight, disappearing and reappearing like mirages as the dolphins dived and surfaced.

Bacher, tall and slender, with light brown hair fastened in a ponytail, scanned the dolphins with her binoculars.

"It's a good-size group," she said. "But I'm not sure about individuals."

Connor turned the boat's wheel over to Kopps and focused his binoculars on the dolphins.

"Ha! Oh, yes, it is the Primas!" Connor said, using his nickname for the Prima Donnas. Connor used moments like this to teach, telling the women about particular features that would help them recognize the Primas in the future. "They're one of the groups I hoped we would see today."

Kopps picked up the boat's speed, setting us bounding toward the dolphins. When we were about a quarter mile away, she slowed to match the dolphins' speed.

The dolphins seemed unperturbed by our arrival and by Kopps maneuvers to keep us in their midst. Connor leaned over the rail and let out a high-pitched whoop that sounded half dolphin and half rodeo cowboy, and one of the dolphins turned its head to peer up at him.

"That's my standard greeting to them," Connor explained, "so they'll know it's me. 'Oh, it's that crazy guy in his boat again.'" Connor liked imagining what the dolphins might say and used a high-pitched, nasal intonation, like teenagers at a mall, whenever he spoke for them.

"Okay. We're basically doing a census, figuring out who is here, what they're doing, and updating our photos of their dorsal fins, since these can change."

Bacher and New took the new photos while dolphins surfaced around us. Some swam bunched tightly together, diving and surfacing, or lightly touching their pectoral fins, a gesture that was something like holding hands and that Connor called petting. Others swam farther apart but were clearly part of the overall group, the way that a flock of migrating birds is together, with some members flitting through a tree and others hopping through a bush.

The dolphins did seem to recognize Connor's call and the boat, because they took turns swimming beside us, as if they were including us in their outing. It was such a wet mash-up of splashing fins and tails that I thought figuring out who was who would be a challenge. But Connor didn't have any trouble. As each dolphin's dorsal fin came into view, he fired off a name. Seeing the dolphins' fins, he had told me earlier, was like seeing the faces of old friends.

"There's Wabbit and Ridge," Connor announced, pointing to two dolphins who were surfacing close together. "And that's Prima up front with Natural Tag. Wow—look at that. Prima has a new scar on his peduncle [the stalk of a dolphin's tail]; that was a big shark bite! Be sure you get a photo of that. There should be seven males. We've known these guys since they were juveniles [about age ten]. They started getting together in 1995 and crystallized as a group in '97. And from the way they're behaving they must have a female with them. Yes. See the dolphin surface in front of Big-Midgie Bites, and between Fred and Ridges? That's Little, a female. So they've got a consortship going on."

We watched as Fred and Ridges swam slightly behind and on either side of Little, flanking her, while directly behind her swam Big-Midgie Bites. The dolphins looked like a playful quartet, surfacing and diving together—except that it wasn't a game, especially not for Little. Connor dropped a hydrophone into the water to pick up the dolphins' calls and handed me the earphones.

"You should hear some loud pops," he said. "That's the sound the males make to keep the females in line."

Through the earphones, I could hear dolphin whistles, squeals, clicks, and squeaks, and then a sudden burst of louder noises—*Popppp; Popppp; Popppp*—like a string of small firecrackers exploding, or, more ominously, like the crack of a whip. In response to the pops, Little slowed her pace, so that she stayed wedged between the three males. Any deviation Little made seemed to elicit another string of pops. After listening a while longer, I set the headphones aside and turned to Connor.

It seemed that Little wasn't here of her own accord.

"Well, yes and no," Connor said. "A female dolphin probably mates with as many males as she can. That confuses the males about who is actually the father of her calf, so they are less likely to harm it.* But if a female is in estrus,

* Female dolphins also have false estrous cycles. These may give a female more time to make the rounds of the various male alliances. By mating with numerous males, a female may be protecting her calf. Although none of the scientists studying the dolphins at Monkey Mia have ever seen a male dolphin kill a calf there, other research teams in Scotland have reported this behavior. Off the coast of California, dolphins have also been spotted hitting and killing harbor porpoises, which are about the size of dolphin calves. The dolphins do not eat the porpoises, and the scientists think these attacks are a kind of practice for infan-

as Little may be, then the males want to keep her with them. The popping sound is like an order. It means, 'Come hither' or 'Stay put.' If she tries to leave, they'll go after her, corral her, jerk their heads at her, maybe bite and slap her."

Male dolphins are wife beaters?

"Well, they can be pretty nasty to the females," Connor said.

Bacher shrugged sympathetically. "I have to say a female dolphin's life here is not easy."

Every four or five years, a female dolphin bears a single calf; for the next two years, she holds no interest for the males. But as soon as her calf is on its own, she becomes the rarest of commodities in the dolphin world: a female in estrus. And every male wants her.

Even as Little swam under heavy guard, she was at risk of being captured by other males—which was why Fred, Ridges, and Big-Midgie Bites were hanging out with their Prima Donna pals. This was a defensive maneuver, a way to ensure that Little stayed with them.

As Connor explained the intricacies of the dolphins' social dynamics, they began to look more like gangs of plotting Lotharios than merry bands of Flippers. He described the Prima Donnas as a "classic second-order alliance"—basically an alliance made up of the other, smaller alliances, "sort of like the Mafia, or a protection racket." Fred and his two pals formed one of these smaller partnerships; Prima, Natural Tag, and Wabbit another. Then there was Barney, who'd once been Fred's best buddy but had somehow lost his status. He still tagged along after the other dolphins, but these days "he's just the 'odd man out,'" Connor said.

Connor called the male pairs and trios "first-order alliances." Such alliances may last a decade or more; others fall apart as the males choose new buddies. As we traveled among the dolphins, I discovered it was easy to spot these partnerships, since the males in these minigangs behaved like dolphin versions of the Three Musketeers. They were "One for All and All for One"—so much so that they seemed at times to be joined by an invisible cord, moving

---

ticide. When a dolphin mother loses her calf, she soon becomes fertile again and available for mating, perhaps by the males who killed her previous baby.

and diving together in perfect synchrony. They surfaced side by side, exhaled at the same moment, and plunged together beneath the waves as if following a choreographer's score. Once, two males made such perfectly matched dives, their bodies arcing into the green waters like silver scythes, that we caught our breath, and Bacher cried, "Oh! Beautiful!" The strongest bonds are between the males with the most synchronous moves. Males reinforce their tight unions by mimicking each other's calls, particularly their vocal signatures, a sound dolphins make that serves the same purpose as our names—and the contact calls of green-rumped parrotlets. (Intriguingly, the dolphin method of social bonding is similar to our own. We also unconsciously match one another's gestures and speech patterns when making friends.)

Male dolphins also seal their partnerships with sex, Connor said. Whenever we went out on the boat that week, one researcher kept track of the dolphins' behaviors, marking these on a data sheet. There were boxes to check for petting, herding, consorting, hunting, goosing (when one dolphin sticks its rostrum in another's genital area), displaying an erection, and having sex—which, among the males, was surprisingly easy to witness (the researchers, though, have not seen a male and female mate). I got used to hearing the researchers call out, "Erection!"—and to seeing a male flop on his back, and another swim on top of him for a quick bout.

"Yes, dolphins are the bonobos of the sea," Connor said, referring to the apes that are well known (or notorious, depending on your point of view) for their "make love, not war" sociability. In bonobo society, everyone seems to have sex with everyone else: females with females; females with males; males with males; even youngsters enjoy a quick rub. The females control the bonobos' hierarchy—and so use sex as the male dolphins do, engaging in genital-to-genital rubbing with each other to form unbreakable alliances, and to relieve social tensions. After all, what better way to flaunt your friendship than with a public display of happy, consensual sex?

Most often the male pairs and triplets travel on their own, away from their larger alliance, as they search for females to corner and keep. If they capture a female, as Fred, Ridges, and Big-Midgie Bites did, and another pair or trio of males from a different alliance rushes in to try to take her away, the first pair summon their allies to help them fight.

As for those nicks, cuts, and scrapes on the dolphins' dorsal fins? They were dolphin-versus-dolphin battle scars, Connor explained. Some dolphins also had scars from shark bites. It was easy to tell the difference between a dolphin-inflicted battle scar and one left by a shark. The latter were large, eye-popping features, like the missing chunk of flesh from Prima's peduncle or the one that had led Connor to name a particular female "Swims With Sharks." A shark's teeth had neatly razored off more than half of her dorsal fin. Dolphins' teeth don't cause the kind of damage that makes you wonder how the victim ever survived, but males and females often emerge from dolphin battles with bloody tooth-rake marks.*

Dolphins fight each other? I asked, feeling that I needed Connor to repeat what he had said—although, of course, if the males slapped their females around, there was no reason they wouldn't go rostrum-a-rostrum with each other, too.

"*Ohhh*, yes, dolphins fight," Connor said. "Big battles. If we're lucky, you'll see one while you're here. I saw my first one on August 19, 1987. And the reason I can remember that exact date is because it was the most exciting day in the history of our project. It's when we discovered the second-order alliances. And it's when I finally began to understand why dolphins have to be smart." His insight came to him in a "flash," he said, and it has changed forever our understanding of the dolphin mind.

On that August morning, Connor recounted as we motored after the dolphins, he and Smolker and two other scientists were watching a trio of males—Snubnose, Sicklefin, and Bibi—as they herded the female Holeyfin close to shore. She had recently weaned a daughter and was probably ready to become pregnant again. The three males kept close beside her, Connor recalled, and dived beneath her at times to nuzzle her genitals. Suddenly, almost stealthily, two other male dolphins—Trips and Bite—swam into Monkey Mia's shallow waters. "Trips and Bite were part of a trio that included Cetus," Connor said, "but Cetus wasn't with them. The two weren't doing much, just hanging there side by side, checking out the other four dolphins. And then they left."

* Dolphins often survive a shark's bite to their backs, but not to their bellies.

Trips and Bite took off at such high speed, Connor was certain that something was up. He had a boat that season, and he hopped aboard to chase after them. About half a mile from shore, Trips and Bite found Cetus, as well as another pair of males, Real Notch and Hi, who were herding another female. The six dolphins milled about for a few minutes, then turned back toward shore, picking up speed as they charged the four other dolphins.

"My heart was pounding," said Connor, who pursued the speeding dolphins in his boat. "I was right on their tails, and I picked up my walkie-talkie and called to Rachel, 'Look out, here we come!'"

Like a silver tsunami, the dolphins surged into shore six abreast. Growling and screaming, they chased the four dolphins through the shallows. The battle—a furious scene of thrashing white water, fins, and tails—lasted several minutes. At the end, Trips, Bite, and Cetus emerged from the frothing waters together with Holeyfin, the female. With the help of Real Notch and Hi, they had won the fight. Most surprising to the scientists, the male dolphins were all friends, not relatives. Yet these unrelated males had worked together to steal a female. Scientists had never before seen genetically unrelated individuals cooperating like this—in dolphins or in any other species, humans aside.

Animals that are not genetically related rarely cooperate. And while it is common in many primate species, such as chimpanzees and baboons, for individual males to join into a single gang to attack rivals, they have never been seen asking for help from another coalition.

For Connor and Smolker, the battle for Holeyfin proved to be an epiphany, the moment when the team's data collecting—all those lists of who was who in the Monkey Mia dolphin kingdom, and which dolphins spent time together, and what they were doing—finally began to make sense. Early on, Smolker realized that the male dolphins had very specific relationships with one another. If she spotted Real Notch, she would soon see Hi. Or if she saw Trips, she was certain to see Cetus and Bite. The females didn't seem to form such tight groups, although they did have loosely bonded alliances and sometimes worked together to chase away males they didn't want to be with.

"That was one of our biggest questions: Why do the males form these partnerships?" Connor said. "In most species, males are rivals. They compete with each other for females. Males don't join with other males; they avoid each other."

Fully grown male elk, for instance, often roam independently until they're sufficiently large and strong enough to drive away other bulls, and to attract a herd of females. Sometimes, two males may fight in a winner-takes-all-the-females battle. There isn't a shred of cooperation between the bulls. And why should they help each other? In all probability, the two are not related. From an evolutionary biology standpoint, there is nothing to be gained by helping some other unrelated guy pass on his genes.

In the battle for Holeyfin, the male dolphins had not behaved at all as male elk do—and Connor immediately recognized the difference. "In almost any other species, all of the males would have been rivals, and would have been fighting," he said. "But they weren't." They had cooperated to battle their foes. "It takes brains to do what the males did, and that means that dolphins are socially smart. It's the demands of social cognition that have driven dolphin intelligence," Connor said.

Despite his sudden insight, Connor did not step onto the world's scientific stage to proclaim that he had solved the mystery of dolphin cognition. He and Smolker had witnessed just one battle between assorted male dolphins, and in that fight had seen a trio of dolphins apparently receive help from another pair. By itself, the event, as heart pounding as it was for Connor, was merely an anecdote; the story of dolphin males teaming up like humans to help each other, nothing more than cocktail chatter. He and Smolker now had to do the hard work that would prove that the Monkey Mia bottlenose dolphins were thinking strategically about social matters—and that doing so had endowed them with unusually large brains.

SEEING MALE ANIMALS COOPERATE is exceptional—so exceptional that scientists created a special category of thinking to explain it: social cognition. The field grew out of two independently written papers published a decade apart (one in 1966, the other in 1976). The scientists who wrote the papers—

Alison Jolly and Nicholas Humphrey—were both attempting to explain the roots and evolution of human intellect, and the expansion of the human brain. Neither was convinced by the prevailing theories that human intelligence was due primarily to being inventive—to creating and making tools and artifacts and devising new ways to hunt or navigate. As Humphrey pointed out in his paper, those skills probably did improve the survival chances of early humans and other primates, such as chimpanzees, that also make and use rudimentary tools. But mere inventiveness wouldn't explain what Humphrey characterized as a chimpanzee's "feats of intelligence" in laboratory settings. There isn't a ready correlation between a chimpanzee's skill at making a tool for fishing termites from the ground and a chimpanzee's ability to recognize its face in a mirror, he pointed out. Knowing that it's your face in the mirror is not something that's going to make life better for a chimpanzee—or any animal for that matter—in its natural environment.

"Why then," Humphrey asked, "do the higher primates need to be as clever as they are?" Jolly asked much the same question, and the two scientists came up with a nearly identical answer: primates (and, by extension, humans) live in such socially complex societies, groupings fraught with so much uncertainty, that members must become "calculating beings" if they are to succeed. "They must be able to calculate the consequences of their own behaviour, to calculate the likely behaviour of others, to calculate the balance of advantage and loss—and all this in a context where the evidence on which their calculations are based is ephemeral, ambiguous and liable to change, not least as a consequence of their own actions. In such a situation, 'social skill' goes hand in hand with intellect, and here at last the intellectual faculties required are of the highest order." Humphrey and Jolly agreed that the most demanding challenge facing primates, our early ancestors, and humans today is the game of social plot and counterplot.

Soon primatologists were busy compiling evidence of Shakespearean dramas in numerous ape and monkey societies, both captive and wild. By recording the primates' most common gestures, such as eye flicks between two friendly baboons, scientists began to understand how primates reach decisions and communicate their intentions to each other. A baboon who wants his friend to help him attack an enemy, for instance, will look at his friend, then at the enemy, then back at his friend, telegraphing his thoughts with his eyes.

Often the researchers noted that the monkeys and apes tried to hide what they were doing and schemed with others like political opportunists in Congress. It didn't take long for scientists to adopt the language of political pundits: their study animals were said to be "deceitful" and "cunning," even "Machiavellian." Here, the descriptions left no doubt, were minds at work.

Social cognition, however, isn't just about being Machiavellian. It is also about learning how to make friends and to cooperate, and as the accumulating research began to show, monkeys and apes are socially smart enough to also be good-natured. They know how to make and keep friends.

To show that dolphins have the same aptitude, Connor and Smolker followed more than three hundred known dolphins, both males and females, for seven years. Conner was well versed in primatology, having elected to study for his doctorate at the University of Michigan with the noted chimpanzee expert Richard Wrangham, even though Connor was the only student who was studying dolphins. Yet he fit right in because primates and dolphins are "more alike behaviorally than you think," he said. He first recognized the similarity in the two species' behaviors in the summer of 1986 as he watched a male dolphin trying to decide between going to the beach for a fish handout or sticking with his female. The dolphin looked toward the beach, then back at her, then back at the beach.

"Fish or her. Fish or her," Connor said, imitating the dolphin's thoughts. "And I realized, because I'd read all the primate literature, that he was acting like one of Jane Goodall's males who once was trying to decide between going to the banana station or going off with his female." He had another epiphany: "Dolphins are herding their females, just like chimpanzees do." But the difference is that dolphin males generally work together to corral the female, while chimpanzees do not. The dolphins need partners for this job for a very simple reason: the females at Shark Bay are roughly the same size as the males, a little more than six feet in length. Thus, a male can't easily boss around a female on his own. He needs a buddy or two to capture her, and he needs them to keep her away from other males, since during the mating season on average there is only one female in estrus for every three males.

"That sets the stage for everything else," Connor said, because it creates a basic conflict for the males. On the one hand, they're competing with each other for females. But on the other, they need each other to win a female. "So they're dependent on each other, too."

There is one more complicating factor: the male dolphins' partnerships aren't always reliable. They operate on a quid pro quo basis: If you help us today, we'll stand by you tomorrow . . . maybe.

"That's where the real social strategizing comes in," Connor said. " 'What did Harry and Jack do with Tom and Bill yesterday?' 'Can we count on them to go after those other guys tomorrow?' Those are the kinds of issues male dolphins face daily, actually hourly." The reward for living in such a socially complex and uncertain society? A large brain.

In 1992, Connor and Smolker published their findings that dolphins, too, can play the game of social plot and counterplot in a major paper in the *Proceedings of the National Academy of Sciences*.* Theirs was one of the first studies to extend social cognition to an animal outside the primate realm—and helped open the door to a more comparative approach to the evolution of social thinking. These days, animals from crows and jays to elephants and wild dogs and even the queens of social wasps are regarded as having social intelligence. "We see the same pattern wherever we look," said Sean O'Donnell, a behavioral ecologist at Drexel University who discovered that wasp queens, which are like colony managers, have significantly larger central processing regions in their brains than do the workers. "The demands of thinking socially lead the queens to have better developed brains."

Connor's and Smolker's paper was highly praised and covered in newspapers around the world, since it cast a new and unexpected light on dolphins and their intelligence. Since then, Connor has worked to fill in the details of dolphins' social intelligence (Smolker left the project in 1994, although the two continued to write papers together). How does a young male choose his partner? What signals and vocalizations do the dolphins use to form their fast friendships, or to call their allies for help? And how to explain the superalliances—the alliances of alliances—that Connor first noticed in 2001? The

* Andrew F. Richards was also an author on this study.

data he and his students were collecting during my week at Monkey Mia would help provide the answers.

EVERYWHERE WE TRAVELED in Shark Bay there were dolphins, males and females, whose dorsal fins triggered memories and stories, and interesting bits of dolphin lore. On several occasions, we watched dolphins fishing in groups. They sometimes work together to herd fish, but each dolphin is responsible for catching his own food, surfacing to swallow his prey whole. Sometimes a dolphin seemed to hold up the fish for his friends to see. This wasn't about sharing but about showing off, Connor said, adopting his dolphin voice to explain what he imagined such a display to mean: "*Nyah, nyah. I have a fish and you don't.*"

Once, we saw two mothers tending their calves in the shallows near shore. The babies had short, rosy-hued rostrums, as if they'd dipped their beaks in strawberry milkshakes. One discovered a blowfish on the sea floor and brought it to the surface to toss in the air. I thought the calves might play a kind of ball game with the fish, tossing it back and forth, but Connor said, "Nope. It's more like a '*See what I have and you don't?!*'"

Dolphins may be cooperative, and able to work together on some goals (such as winning a female), but they do not share. They also do not steal.

"It seems to be a rule," Connor said. "You don't try to eat somebody else's food. Dolphins are polite about that. Polite, but selfish."

Another day we came upon Seven, a male in the Blues Brothers' alliance. He had a round, orange puff on his beak that looked like a weird growth—but it was a sponge he had intentionally picked up on the seafloor. He had slipped it over his rostrum as we would put a glove on our hand. Thus equipped, he could forage in the rough sediments for burrowing fish. Smolker had discovered the dolphins' use of sponges—the first evidence in the wild that dolphins use tools.

"Usually, it's the females who carry the sponges, since they invented this technique," Connor said, something the scientists realized after observing that females were the primary spongers. "Sometimes their sons learn to do it, like Seven has. It's very clever," Connor agreed, "but using tools isn't what led to dolphins' big brains." Sea otters make equally remarkable tools, he points out, but they don't have large brains. Coming up with new ways to put dinner on the table is simply not as mentally taxing as juggling social affairs.

Just how complicated the dolphins' social lives can be (and how tricky it is for the scientists to figure out what the dolphins are doing) was driven home on my last morning at Monkey Mia. Every day, Connor had hoped that we would see some of the dolphin alliances join in battle. He needed more observations of these fights, and he wanted to explain one in detail to me as it unfolded. But all the dolphin gatherings we encountered had been peaceful. Soon after we set out on the bay on this day, however, Connor spotted some members of the Kroker Spaniels, one of the larger alliances. Even from a distance, he could tell that they were slightly agitated and feisty. When we motored up alongside, we found six males herding two females. Connor gave his rodeo hoot, "*Eeeeyippee-yai-yai!*" and then pointed to another trio of dolphins heading our way. "It's more of the Krokers. Get ready for a fusion," he called to New, who held the data sheet.

Leaping over the water's mirrored surface came Ceebie, Moggy, and Pasta. They dived in among the other dolphins, and all joined in a session of splashing, rubbing, and petting. There were now nine males and two females.

"They're going to want to get a female for their buddies," Connor predicted. "Listen," he added, dropping the hydrophone into the water and handing me the headset. "They're planning on going places and doing things."

Dolphin chatter filled my ears—an underwater symphony of whistles, squeaks, clicks, growls, and the inevitable pops for ordering the females about. Connor and other researchers think that noisy sessions like these may be the dolphins' way of deciding what they're going to do next. Apparently, the Kroker Spaniels came to an agreement, because they began swimming together in one direction and were soon picking up speed.

"Okay, they are accelerating," Connor said, as we followed along in their wake, traveling so fast that we could no longer use the hydrophone.

Three of the dolphins in the lead hurtled out of the water, throwing their bodies through the air like spears to travel even faster.

"See those other dolphins up ahead," Connor shouted, pointing to several other fins on the horizon. "That's where they're aiming. I bet that group has a female and these guys want her."

Connor sped ahead of the Krokers to find out which dolphins they were targeting. It was Morsel and Edge, and in fact they did have a female with them, named New Fin. "They don't stand a chance against the Krokers," Connor said.

But just then, another group of three males arrived from a different direction—the allies! Morsel and Edge churned together through the water with their three homeboys, making shallow dives, sometimes flipping backward, rubbing their faces and fins together while the Krokers continued to speed our way.

"Get ready with the video camera!" Connor called to Kopps. "There's going to be a fight."

The six dolphins we were with heaved about in the water, agitated by the sounds of the approaching Krokers, who, Connor thought, were probably calling among themselves to keep their members on track. Then, suddenly, the three new males—the allies—sped away from Morsel and Edge, taking New Fin, the female, with them. Would Morsel and Edge have to face the Krokers' onslaught alone?

We watched the advancing Krokers race to a . . . halt. They stopped some distance from Morsel and Edge, lining up in two rows with their pectoral fins splayed at their sides, a dolphin's resting pose.

"What happened, guys?" Connor called out to them. Then he laughed. "Well, I don't know what that was all about. But Morsel and Edge have lost their female."

The two dolphins swam forlornly together. Had their buddies betrayed them or protected them from a serious battle by swimming away with New Fin?

Connor shook his head. "We've been watching these dolphins for almost twenty-five years, and we still see new things. There are better technologies now—new hydrophone arrays—that we want to use to investigate their vocalizations, which could help interpret events like this."

This was another project for the future.

We slowly motored back to join the Krokers, who were now diving happily for fish. Stealing a third female for a grand ménage seemed the farthest thing from their minds.

"There's one thing my work doesn't show," Connor told me later. "It doesn't show that dolphins have theory of mind," the ability to understand what another individual is thinking.

One of the biggest debates in animal cognition is about whether any species, aside from humans, is capable of recognizing that another being has a mind. Some of Nicola Clayton's experiments with blue jays and other researchers' studies with chimpanzees strongly suggest that both species share this aptitude, and many dolphin and other cetacean researchers expect that their species do as well—an opinion that Connor shares, even if he lacks the hard evidence.

"Dolphins couldn't do all the things they do with their alliances if they didn't have theory of mind. If you're thinking strategically in social situations, as dolphins are, you need to know how to read others' intentions. You need to know that they have minds."

JOHN LILLY THOUGHT he could fathom the dolphin's mind by bonding with them and using interspecies communication. In his two most popular books, *Man and Dolphin* and *The Mind of the Dolphin*, he recounts his efforts. At the time, the idea of theory of mind—recognizing that other individuals have minds and can think—had not yet been invented. The ideas behind social cognition, too, were not yet fully formed. But Lilly thought the dolphins had minds and language and that they had something to communicate to us.

In 1965, Lilly devised his most elaborate experiment for breaking the "language" barrier between dolphins and humans. (I put "language" in quotes because scientists have not found any evidence that dolphins have a language as we know it. Connor and others suspect that dolphins may be able to link friendly or hostile sounds together with another dolphin's signature contact call to convey their feelings or intent, but this has yet to be proved.) Borrowing from foreign-language schools, Lilly decided that the best way for dolphins to learn human speech was through total immersion—meaning that they should live only with people, not dolphins, and hear only people, not other dolphins, speak. In his dolphin research facility on the island of Saint Thomas, he built a special flooded house where a person and a dolphin could live together.

The person selected for this experiment was a dark-haired, pretty young woman, Margaret Howe, who had been working as a hostess in a restaurant on the island. She caught the eye of Lilly's good friend the astronomer Carl Sagan, who tried to impress and seduce her—partly, it seems, by squiring her around Lilly's lab to show her the dolphins. There she met the anthropologist Gregory Bateson, who was then managing the lab and experiments. Sagan's romantic plot didn't work out as he had hoped: Howe rebuffed his advances. But she was intrigued by the dolphin research, and Bateson hired her to carry out Lilly's immersion experiment. Lilly had wanted someone who wasn't trained as a scientist, he wrote in *The Mind of the Dolphin*, and who had no preconceptions about dolphins, and Howe fit the bill. She was also a good observer and writer, and it probably didn't hurt that she was attractive.

For two and a half months, Howe lived in the flooded house with a young adolescent male dolphin, Peter, one of the former stars of the television show *Flipper*. The "house," as Lilly called the enclosure, was basically a wide, shallow pool that held twenty-two inches of seawater and came equipped with an air mattress (usually damp) for sleeping, freshwater shower, desk and chair, and telephone. Before moving into the house, Peter had lived in the lab's deep tank with two other dolphins, both females. Howe had met him there and learned how to give him his English language lessons.

In the house, Howe had a strict schedule to follow. She was to rise at 7:30 each morning, feed Peter five pounds of fish between 8:00 and 8:30, and give Peter lessons and play with him for half-hour sessions until 6:30 at night. There were lunch, rest, and dinner breaks, too. But even during those times, Lilly instructed Howe, she was always to be aware that she was living with Peter. Human and dolphin went to sleep at 10:00 p.m., Howe on her wet mattress, and Peter lying in the water close to her.

Howe worked on teaching Peter to count in English and teaching him the words for various shapes. Dolphins use their blowholes to mimic human sounds, so Peter often sat with his head out of the water, or just slightly submerged, next to Howe. There are photographs of Peter looking up at her intently as she leans forward on her metal chair and speaks to him or talks on her phone. She wears a black leotard, and the water laps at the seat of her chair. Other photos show her in the water with Peter, sometimes with her arms wrapped around his torso, other times with him swimming through her legs—which she didn't like because his hard pectoral fins bruised her shins.

After two weeks, Howe was able to report that Peter was making good progress on his "humanoid" speech. He could count to three, could pronounce "ball" clearly, and was practicing saying the first letter of her name, "M." He pretended to chat in these new sounds, inventing his own variations, when she talked with her friends on the telephone. Most troubling for the human side of this partnership (but probably not for the sociable dolphin), Peter never left her alone. "*He does not go away,*" Howe wrote in some alarm in her notes. Peter was "more like a shadow than a roommate." He was "continually" at her feet, touching, pushing, nibbling, speaking, squirting. "He does not get distracted or bored, *he stays right with you.*"

Peter's nibbles could hurt, and Howe resorted to carrying a broom to fend him off. She didn't want to whack him with the broom, or shove him away, and he apparently didn't want to hurt her, because he would become gentle when she shouted in pain. To show he meant no harm, Peter would turn on his back and lie still.

All of Peter's nibbles were a kind of dolphin courtship, but he couldn't get very far because Howe was afraid of his teeth—something that Peter recognized, Howe wrote. So Peter "devised a subtle, gentle method" to overcome her fears. He picked up a small ball and held it between his teeth as he approached her—a sign he meant no harm, because he could not nip her while holding the ball. He then nuzzled her feet and legs, while watching her face. As she relaxed, he ever so slowly let the ball slip farther back in his mouth and then drop into the water—making it possible for him once again to take her leg in his mouth, but gently, she wrote, without hurting her.

Howe knew that Peter was wooing her. For most of their two and a half months together, he worked at teaching her to "trust him." Howe says she was "flattered" by Peter's courtship and touched by the patience and care he took to win her. And as she relaxed, Peter did, too, turning on his side and back for her to stroke his belly and genital area. Although male dolphins' penises are tucked away in the genital slit, they emerge when erect. Peter had erections, and ultimately, Howe did what Peter wanted: she rubbed his penis for him, bringing him to orgasm. It was the only way, she found, that he could relax and pay attention to his lessons.

There are of course endless jokes one could make about the new levels of interspecies communication Howe and Peter achieved. But Howe's account is more poignant than titillating. Here was a young dolphin who had probably spent half his life in captivity. He was a sexual being with a dolphin's urges, and he should have been bonding with a gang of other young males. Instead, he found himself alone in a pool with a foreign creature—an alien—who was afraid of him. And Peter taught her to trust him. Did Peter and Howe use the positive-negative reinforcement technique of the behaviorists to reach their understanding? Or could it have been that they were reading—and responding to—each other's minds?

9

WHAT IT MEANS TO BE A CHIMPANZEE

Chimpanzees bridge the gap between "us" and "them."

JANE GOODALL

No one knows where Keo, a chimpanzee at Chicago's Lincoln Park Zoo, came from, other than somewhere in Africa, or who his parents were, or whether he had brothers and sisters, or aunts and uncles and cousins.

"But he probably did, since chimpanzees live in extended families," said Stephen Ross, who supervises the zoo's great ape cognitive research studies. "All we know is that Keo came to us as a one-year-old toddler in 1959. He's been with us ever since."

Keo, who's now in his fifties, lives just a short walk from Ross's small office in the zoo's Regenstein Center for African Apes, but I had not met him yet. There were other matters that needed Ross's attention first, specifically two seven-year-old male chimpanzees that a woman in Delaware had raised but could no longer keep. Ross was trying to find them a safe haven, he had explained to me when he greeted me at the center's main door earlier that morning. The zoo had not opened for the day, and he wanted to give me a tour of the building, which he had helped design, while it was still quiet. But Ross was also in the middle of rescue negotiations for the chimpanzees. He referred to them as the "boys in blue denim," because of a photograph their owner had sent him of the two apes dressed in matching denim shorts. He didn't laugh at the name or the photo; he wasn't derisive of the owner or her love for her boys. All he wanted was to get them out of their dress-ups and into an environment where they could be chimpanzees again.

Ross is in his late thirties, athletic and handsome, with dark eyes and the pallor of an office worker. Officially, he is a primatologist and the assistant director at the zoo's Lester E. Fisher Center for the Study and Conservation of Apes—although neither title adequately conveys his passion, which might best be summarized as trying to understand what great apes, and chimpanzees in particular, are thinking about, so that people can provide a better life for those in captivity. He was dressed for his dual roles in blue jeans and a striped long-sleeved shirt, the cuffs, even at this early morning hour, rolled back.

Because scientists have studied chimpanzees for more than a hundred years, Ross can—and does—draw from a wealth of research about the chimpanzee mind, some of which he incorporated in the design of the new center. He and his colleague primatologist Elizabeth Lonsdorf, who is director of the Fisher Center, also have projects under way in the wild and at the zoo that will help us understand how chimpanzees perceive and categorize their world, what's important to them, and how their mental capacities stack up against those of other great apes, including humans. Making arrangements to move chimpanzees from a private home to a public zoo is not really part of Ross's job description, but he's also the chairman of the Chimpanzee Species Survival Plan for the Association of Zoos and Aquariums. The program attempts to ensure that the three hundred or so chimpanzees in zoos across the United States are well treated, not bred to close relatives, and not alone, but housed with other, compatible chimpanzees.* "I think that's the number one, most important thing to every chimpanzee: having a social relationship," Ross said.

Ross had come to understand the social needs of chimpanzees when working in the 1990s as a research assistant to Mollie Bloomsmith, first at a

* In 2009, Ross also launched ChimpCARE, an organization committed to improving the well-being of the two thousand chimpanzees in captivity (zoos, research labs, private homes, and sanctuaries) in the United States. There are three times as many chimpanzees in captivity in the United States than are now found in the wild in Tanzania.

medical lab at the University of Texas and then at Yerkes National Primate Research Center near Atlanta, where she was the director of the behavioral management unit. In both labs, chimpanzees are used in biomedical research. When Bloomsmith and Ross arrived at Yerkes, nine male chimpanzees were living solitarily. Bloomsmith wanted to show the center's staff that the chimpanzees would be better off living with companions; Ross's job was to figure out how to introduce the chimpanzees to one another without causing mayhem. In the wild, male chimpanzees attack unfamiliar chimpanzees. They do so in captivity as well unless they are slowly introduced and eased into a friendly relationship. But because chimpanzees yearn for companionship, Ross said, even those that were once strangers can eventually be housed together. Ross spent three years helping the nine chimpanzees make friends, and when he left, only one was still living alone. "I was worn out by then emotionally," he said, "but I still feel that I could have—that I should have—done more."

Ross chairs the Chimpanzee Species Survival Program partly to assuage what he calls his "Schindler's List" feelings. The program isn't designed to rescue chimpanzees in private homes, or in medical research labs; it works to improve the lives of chimpanzees in zoos across the United States. Still, Ross gets calls, like this one, from people seeking new quarters for their pet chimpanzees—and even though he knows that most zoos don't have room for more chimps, and that chimp sanctuaries are full and at a crisis stage, and that asking the Little Rock Zoo to take these two adolescents would be stressful and an enormous burden for the workers there ("The staff will have to be with these chimps every night for months, even years"), he does everything he can to help.

"I never wanted to be the next Jane Goodall, following chimps in the wild," Ross said, although it was her classic book *In the Shadow of Man* that inspired him as a twelve-year-old living in Ottawa, Canada, to devote his life to chimpanzees. "I've always wanted to look after chimpanzees in captivity." He wasn't sure where the desire came from; he'd never seen a live chimpanzee until he was twenty-four, when he took the position at the University of Texas (and was then surprised at how large they were). He simply felt an affinity with the animals Goodall described and wanted to help make life better for those in zoos. "I didn't know exactly how I was going to do that," Ross said, "but I figured I would need to know something about biology and caring for animals."

So he earned his biology degree at the University of Guelph and worked on a project in the summers to improve the housing of domesticated pigs. When he heard that the scientists who manage a research colony of rhesus macaque monkeys on Cayo Santiago Island, Puerto Rico, needed a volunteer research assistant, he jumped at the chance. It was his first experience working with any primate, and he spent a year following and recording the monkeys' daily activities. From Cayo Santiago he moved to Texas, and then to Yerkes. And there he met a young chimpanzee named Drew.

Sometimes, a scientist is so strongly affected by a single animal that the encounter changes him or her forever. Jane Goodall had such an experience with David Graybeard, the first chimpanzee at Gombe to tolerate her presence, and Irene Pepperberg lost her heart to Alex the gray parrot, despite her determination not to. Drew had a similar effect on Ross.

Drew's photograph hangs on Ross's office wall; when he looks up from his desk, his eyes meet those of Drew. "He represents all that's good about chimpanzees," Ross said. "He was gentle, thoughtful, and he could relate to humans as well as other chimpanzees. We had a strong emotional connection." Did that mean that Drew loved him? Ross nodded. "Or maybe I loved him."

At Yerkes, Ross installed old tires, climbing bars, and towers for the chimpanzees to play with. Drew liked these, but he especially enjoyed a video task that Ross and Bloomsmith designed as part of their chimpanzee welfare study. To earn treats, Drew had to maneuver a joystick in response to what he saw on the monitor. "He wasn't particularly good at it," Ross said, "but he loved it because it exercised his mind, and it was something we did together." After Ross left, though, none of the staff had the time to continue playing Drew's favorite game.

Ross returned to Yerkes six years later, and by then Drew had grown into a disgruntled, feces-throwing, cage-pounding adult. "The staff insisted I put on a protective suit before going in to see him, something I'd never had to do before," Ross said. "He was banging his cage, and I shouted, 'Hey Drew,

I'm coming in there. Get ready. I'm coming in.' And all the banging stopped. He was absolutely silent. I think—I know—he recognized my voice. When I turned the corner, he was waiting for me, right where we used to do the joystick game together. That was a defining moment for me, and I cried. You could see the surprise on his face. Also the disappointment, and confusion."

Ross took a deep breath, and studied Drew's photo before continuing. "Drew had a massive influence on me. I felt I had failed him, and I wasn't going to do that to a chimpanzee again. Many of my ideas about designing this building and making a better life for the chimpanzees here are because of Drew."

Ross's tenure at Yerkes had also taught him what he considers the most important lesson about caring for chimpanzees: "*They're not you.* I had to forget about what *I* thought would interest them, and watch them. That's how you learn what they need to be fulfilled. Their interests and behaviors tell you."

Ross knew what it meant to love and to lose a chimpanzee, and he felt the anguish of the woman with the two youngsters; he also knew what her chimpanzees most needed. He had expected her to say this morning that she had agreed to his plan. Instead, she was asking for more time to think about what she should do.

"Well, let's go meet Keo," he said, standing up. "He's waiting for me to say good morning to him anyway."

From Ross's office we followed a corridor to Keo's home, which he shares with three other chimpanzees: his forty-two-year-old daughter, June; another female, also in her forties, named Vicky; and Vicky's twenty-eight-year-old daughter, Kibali. June's mother lives at a different zoo. Keo had recently turned fifty, which is the chimpanzee equivalent of being almost a one-hundred-year-old human. For the first forty-four years of his life at the zoo, Keo had been kept in a glassed, indoor enclosure. When he was a youngster, he had been used as a clown and trained to take part in children's tea parties. He wore a jacket and party hat, poured tea from a teapot, and sipped it from a teacup, while properly holding his pinky aloft. If you gave him a teacup now, he would hold it that same way, Ross said. Later, once Keo was an adult, he was placed in a cage and on exhibit, a chimpanzee to be stared at.

During all those years, Keo never smelled the earth after the rain or felt the sun warm his back. Only after the zoo opened this center in 2004 did Keo step outside into the open air. He was still confined, of course, in a fortified arena, surrounded by tall glass windows, but the arena had two parts to it, a spacious indoor area, with poles and ropes for climbing, and a large, grassy space outside that, with bamboo thickets, faux trees, ropes, and a waterfall to explore—providing at least a semblance of the wild. The four chimpanzees could stay inside or enjoy the air outside as they chose, something that made for far happier apes, Ross said. Right now, the public could see these four only if the chimps were outside. There were seven additional chimpanzees and twelve western lowland gorillas for visitors to view in the main part of the center. It, too, was designed to the give the apes a simulated forest for a home, with ladders, poles, and platforms to climb, and the choice to travel between the indoors and outdoors.*

"You could see they liked it here from day one," said Ross, who has been at the zoo since 2000. While here, he'd also studied at the University of Copenhagen and had earned a doctorate in primate management, behavior, and welfare. "They do a lot less of that idle scratching that apes do when they're stressed."

On this sunny morning, Keo was still inside. He ambled over—upright, on two legs, his preferred walking style—to the window to greet Ross and me. Only the silvery beard below Keo's chin hinted at his age. The rest of his fur was a lustrous black. He nodded his head at Ross and loosely tossed his hand in the air to wave hello, a human greeting he had learned as a youngster from his trainers.

Keo was less certain of me. We were separated from each other only by the wall of glass and could have looked directly into each other's eyes if I'd leaned down a bit. Many primates (humans included) consider staring to be a threat, so I gave him brief sideways glances while jotting my notes. Keo, though, who had dropped to all fours, was looking me up and down, although

* This is an increasingly common way to house great apes and is also used at such places as Chimp Haven, the National Chimpanzee Sanctuary outside Shreveport, Louisiana, and, during the summer, at Germany's Leipzig Zoo.

with the slightly bored, put-upon expression of a village elder who has seen one too many anthropologists. At last, he began to turn away, when, to my surprise, he sidled closer—not to look into my eyes, but to peek at my notebook. Instinctively, I moved it into a lower position and held it at an angle so that he could watch as I wrote.

"You want to take a look?" I asked. "I'm afraid my handwriting is pretty sloppy."

Ross smiled. He'd been standing with his back against the wall, one hand shoved in his jean's pocket, but leaned down to engage Keo. They both nodded rapidly at one another, a kind of chimpanzee *you're-okay-I'm-okay* greeting.

"He's trying to figure out what your relationship is to me, and if he should be jealous. I think he's decided that he doesn't need to be," Ross said. "He can feel threatened by strangers, especially men with facial hair. If he was jealous, he would hit at me—hit the glass with his fists and feet. Or he might fill up his mouth with water later, and wait for the right moment, and then spit. I've been doused a few times."

Ross leaned back against the wall. "I think he views me as one of his group, or as a hybrid. I'd like nothing more than to know what he and the others really think about us, how they view us, and their world in general. From observational studies that have been done both in captivity and in the wild, and from what I see with our chimpanzees, I would say that they're most concerned about their social relationships, the hierarchy. That's the one thing that's always on their minds, I think. But it's difficult to get answers to more specific questions about how their minds work through observations alone."

For a finer understanding of chimpanzee thoughts, Ross and other researchers use experiments, although a better term might be "challenge tests," since nothing is done to the chimpanzees. Instead, they're put in a situation and watched to see how or if they can solve it. Chimpanzees have been taking such tests since the early twentieth century, when Wolfgang Koehler showed, among other things, that they were able to figure out how to stack boxes to reach a banana hanging overhead. Researchers today also give chimpanzees physical tests as well as virtual ones that use special, primate-proof touch screens linked to computers. Chimpanzees and other great apes are happy to sit in front of touch-screen computers to play games and take tests if they're rewarded with grapes or other treats, Ross explained.

The zoo's touch-screen computer testing program was in its infancy, but Ross said it had been used elsewhere to gauge chimpanzees' short- and long-term memories and to understand how they perceive their environment, while making it engaging and fun. Ross and Lonsdorf were also exploring the chimpanzees' problem-solving skills by letting them fish for ketchup hidden inside a fake termite nest. The chimps had to find and make their ketchup-catching tools from the bushes in the outdoor arena and then sort out among themselves who got the privilege of dipping his or her stick into the nest. All the experiments were purely voluntary. "They can join in the tests, or not. It's up to them," Ross said. "But they seem to like participating. We give them a signal when one is under way, and they show up when it's their turn."

Ross planned to show me some of the chimpanzees and gorillas at their touch-screen tasks that afternoon. We said good-bye to Keo, who had moved to a corner in his indoor enclosure. It was Keo's favorite spot, Ross said. From there, the elder chimpanzee could keep an eye on the humans coming down the hall, as well as the doings of his own small troop: June, sleeping on a blanket against a wall with her favorite toy, a red rubber duck; and Vicky and Kibali, nestled together in an overhead rope net, grooming each other. A few sparrows had flown inside and were hopping among the wood shavings that covered the floor. The wind was rustling the bushes and bamboo outside, and the sun was shining on the faux bamboo climbing poles and the ropes that dangled among them like fat vines in an African forest somewhere far away.

IN 1960, only a year after Keo arrived at the Lincoln Park Zoo, Jane Goodall set out to make contact with wild chimpanzees in the Gombe Stream National Reserve of Tanzania (called today Gombe Stream National Park). Almost nothing was known about the behavior of chimpanzees in their natural habitat when she began her study, as I noted in the Introduction. In 1890, one American adventurer, Richard L. Garner, had built a cage in a West African forest where he sat daily for several months taking notes on chimpanzees that chanced to walk by. And there were the random reports of travelers who had witnessed chimpanzees using rocks as hammers to break open oil palm nuts, and dipping sticks in beehives to reach the honey.

Goodall's plan was different, she'd explained to me during my 1987 visit. She intended to spend five years documenting the individual chimpanzees and their daily lives in the forest. Louis Leakey, the paleoanthropologist who sponsored her research, thought that such a study would shed light on how our earliest human ancestors had lived. He also thought her project was urgent because he feared expanding human populations would ultimately lead to the extinction of the chimpanzees and the other great apes. Neither Leakey nor Goodall anticipated that her Gombe Stream Research Center, as Goodall titled the site of her study in 1965, would still be going full bore as it is today. Nor did they foresee the radical changes Goodall's observations (and those of other field researchers) would have on our understanding of the chimpanzee mind—and on the study of animal minds overall.

At first, Goodall "couldn't even talk about the chimpanzee mind because chimpanzees didn't have one." She made this quip at a conference held at the Lincoln Park Zoo in 2007 that was titled, ironically, "The Mind of the Chimpanzee." But then, no animal—other than the human animal—was regarded as having a mind in those days, at least not by most Western scientists. Strict behaviorism was in force, although Goodall, who had only had a few months of university training, was largely ignorant of its concepts—and so she broke every one of its rules. She gave the chimpanzees names. She wrote about them as individuals, with distinct personalities and temperaments. She distinguished males from females, young from old, as opposed to lumping them together as a class—"the chimpanzee"—or referring to a single chimpanzee as "it." Instead, Goodall dignified them with the pronouns *he* and *she*, noting that male and female animals behaved differently and had different interests and roles in their societies. She collected notes (which others would at first dismiss as anecdotes) about what and where the chimpanzees ate, whom they groomed, whom they fought with, whom they had sex with, and how they played, cared for their young, patrolled their territories, and hunted and killed monkeys and bush pigs.

Goodall wanted to understand what life is like for a chimpanzee in that natural world, and—at least to herself, in the beginning—she ascribed to them emotions and intelligence enough to make decisions. She did not realize how subversive or revolutionary her approach was until she submitted her first scientific papers. But Goodall kept at it, letting her discoveries speak for themselves—and, with each one, chipped away at the prevailing notion that humans alone are capable of thought.

One of Goodall's greatest breakthroughs occurred only four months after she arrived at Gombe, when she watched her favorite chimpanzee, one she called David Graybeard, "carefully push a long grass stem down into a hole in [a] termite nest. . . . It was obvious that he was actually using a grass stem as a tool." To verify what David was doing, Goodall imitated his actions. After David left the mound, she poked one of his discarded stems into the nest, pulled it out slowly, and found a cluster of termites clinging to its end. Later, Goodall would see the chimps modify twigs and stems, stripping away the leaves to make proper termite- or ant-retrieving tools. In *In the Shadow of Man* she noted that hers was "the first recorded example of a wild animal not merely using an object as a tool, but actually modifying an object and thus showing the crude beginnings of toolmaking."

Goodall's discovery thrilled Louis Leakey, who announced it at a press conference at the National Geographic Society in 1964 by first alerting his audience that scientists needed to revise the definition of the human genus, *Homo*. The previous definition, he said, had described *Homo* as a creature capable of making tools "to a set and regular pattern." But now that Goodall had witnessed chimpanzees doing exactly the same thing, Leakey said, a "new definition of man" was required that would be "very much more complex." He added, "We decided we must exclude chimpanzees from the United Nations."

That same year, Goodall published her "chimpanzee as tool user" discovery in *Nature*. Since then, she and others have documented so many other striking behavioral and mental similarities between humans and chimpanzees that scientists now struggle more with the ethics of studying chimpanzees, especially those used in medical research, than with finding some new definition that will separate us from them.* And they worry about how best to conserve those few small populations of chimpanzees that remain in the wild.

As Goodall recognized, chimpanzees have distinct personalities that affect their success in life; they create cultures, wage wars, and are socially astute. They grieve for their dead, plot coups against their leaders, act kindly toward chimpanzees that are ill or crippled, and display some aspects of theory of mind, understanding what another individual knows or is thinking about. In short, they "manifest intelligent behaviour of the general kind familiar in human beings," as the psychologist Wolfgang Koehler concluded in 1925, "a type of behaviour which counts as specifically human." It required nearly another hundred years of study in the wild and in the lab—and many great debates, some of which are still unresolved—for science in general to reach the same conclusion.

Scientists are not suggesting that chimpanzees are minihumans in fur coats, as Victorian-era artists often depicted them. They are chimpanzees, and it is the chimpanzee mind—not the chimpanzee mind on its way to becoming human—that researchers like Ross and Lonsdorf most want to understand.

WHEN I JOINED ROSS again after lunch, we rode an elevator to the center's basement to watch gorillas and chimpanzees taking touch-screen tests.

If the apes' enclosures upstairs resembled patches of forest, the basement's holding area felt like a boiler room. The floors and walls were made of cement, and the lighting came from fluorescent bulbs that ran overhead, alongside the center's plumbing pipes. The apes were confined to steel-barred cages.

---

* In 2007, the parliament of the Balearic Islands, an autonomous province of Spain, passed the world's first legislation granting legal personhood rights to all great apes. The following year, a parliamentary committee urged the country overall to follow suit and grant the rights to life and liberty to the apes; the committee's resolution has not been enacted. New Zealand started this trend in 1999 by giving the five great apes basic rights. They now can no longer be used in research, testing, or teaching.

We arrived just as Jill Moyse, the gorillas' primary trainer, was summoning them to this nether world. She was dressed in the zoo's standard uniform of khaki pants, kelly green T-shirt, and rubber boots and had pulled her brunette hair into a ponytail. Moyse and Katherine Wagner, an intern, were waiting for the gorillas to climb down the ladder that connected their enclosure upstairs to the basement. Wagner, who wore a white lab coat over her sweater and jeans, would be running and videotaping the touch-screen test.

"Come on, Kwan!" Moyse called, looking up the ladder to see if the group's silverback male was on his way. To us, she said, "Don't look at him. Keep your eyes down."

Kwan's family had four members, three females and a three-year-old male toddler. They would all follow where Kwan led, Moyse said. To a silverback, a direct look is a threat, and if I broke the rule, he might charge the bars, possibly injuring himself. It didn't matter that I couldn't see Kwan and the others. I smelled them—a pungent odor, both sharp and musky, filled the air as the gorillas entered the basement.

Moyse called Kwan to the bars of the cage for a treat, and while he was focused on her I sneaked a peek. I'd been told he weighed three hundred pounds, and all of it looked like muscle. His arms and legs were shaped like barrels, and his shoulders were as broad as a doorjamb and splashed with silvery-tipped hairs—the insignia of male gorilla status and power. His face was inky black, almost metallic in hue, and his brow bulged like a visor over his dark eyes. If he wanted, he could be fearsome. But in Moyse's presence he was docile, opening his mouth wide so that she could pour in a full bottle of Citrucel, both a treat and a health supplement.

"Okay, Kwan," Moyse said, as Kwan smacked his lips.

Without further ado, Kwan climbed up on a bench that faced a 36-inch computer screen. His females lingered nearby, while Amare, the little male, clung to the bars and stared out at us, wide-eyed. Sitting alone at the touch screen, Kwan looked rather lordly. That was one of the differences between gorillas and chimpanzees, Elizabeth Lonsdorf had told me earlier. Although gorillas live in family groups, they tend to do less social grooming than chimpanzees, and the silverback male generally preferred to sit by himself. Lonsdorf had given the gorillas the chance to try fishing for ketchup from the zoo's faux termite mound and had discovered that they enjoyed it, even though they don't fish for ants or termites in the wild. But the gorillas preferred to fish alone, or in mother-and-infant pairs, while the chimpanzees fish in groups and are comfortable crowding together while watching one another's success.

"Kwan has to put four letters—O, X, S, I—in that particular order," Ross said, explaining the test. "The letters don't mean anything; they're just shapes we've chosen for the apes to remember. We're trying to get an idea of how they perceive shapes and if they can remember a sequence of things, which should give us some idea about how their memories work. Very few studies have been done like this with gorillas. Most people think gorillas are dumb—not a lot going on upstairs. The truth is that we don't know what they're capable of. They just haven't been tested as much [as the other species of apes], or in the right way. Kwan is actually good at this."

Wagner stood at her computer, which sat on a chest-high desk, and entered the testing program. Kwan looked intently at his screen as the letters—large, blocky, and colored white—flashed up, sometimes spread widely across the dark background, other times set more closely together. He gently tapped each letter with the back of a knuckled finger. Every time he put the letters in the correct order, Moyse called out an enthusiastic "Good boy, Kwan!" the computer emitted a high-pitched beep, and a grape rolled out of a dispenser for him to eat. If he entered the sequence incorrectly, the screen flashed red and emitted a low note, signaling "wrong."

Amare turned away from us to watch his father's progress as Kwan worked his way through thirty trials in a five-minute period. Out of these, he was correct 67 percent of the time.

"If he was only guessing, he would get 4 percent right," Ross said. "So he's performing well above random chance. Once he gets to the 80 percent level and manages to stay there through three trials, we'll add another symbol."

"There are so many factors that affect their success at these tasks," added Lonsdorf, who had joined us. Her dark hair spiraled in curls to her shoulders, and she had the kind of direct gaze and patient manner of an ethologist who spends months in the field, focused on her study animals. "Kwan takes longer than the chimpanzees because he has to keep an eye on his females and make sure that Amare's behaving," Lonsdorf continued. "But you can't have

Kwan down here alone without his family; silverbacks have to keep an eye on all their members—that's their job. If you didn't know this, or take it into account, you could conclude that gorillas aren't as good as chimpanzees at this test."

After Kwan finished, Moyse directed the gorillas into another holding area, once again calling each one by name.

Then Keo's group of chimps came thundering down the ladder and moved into separate caged areas, where another keeper handed them cups of juice. Compared to the gorillas, the chimpanzees were like hooligans. They hung on the bars, staring to see which humans were around, pounded on the walls, and hooted loudly.

Moyse directed Vicky, the mother of Kibali and the main test taker in Keo's group, into the touch-screen cage, and Wagner set the test in motion. Without hesitation, Vicky set to work, finishing up within three minutes but scoring only 40 percent right.

"Good job, Vicky!" Moyse called to her. Vicky nodded her head but rapped at the screen.

"She wants to do more," Ross said. "Tomorrow, Vicky. You'll have another chance tomorrow. She can do better; she's scored as high as 46 percent, but she's sometimes so frantic to be right she doesn't touch the screen precisely. She just wants that treat; so that's something that also affects their success—how much they can control their impulses."

Keo and the others took their turns at the screen, too, but they were still very much at the trial-and-error stage, Wagner said.

"Keo started off doing great," Ross said. "I think he's really bright, but he lost his motivation. He wasn't that interested in scoring a few grapes or jelly beans. If one of them doesn't want to do it, we drop them; it's their choice."

The real chimpanzee star, Optimus, was in the zoo's other group and would be tested later. But even Optimus had been surpassed by a young female gorilla, Rolli, which initially surprised the researchers.

"Rolli takes just a little bit longer to survey the screen; she isn't in quite as much of a hurry, and she gets the high scores," Ross said. "It's that impulse-control thing again."

Other researchers have found that orangutans, who live more solitary and slower-paced lives than chimpanzees, also surpass chimpanzees at certain tests. At the Leipzig Zoo, German researchers placed a tempting banana behind a Plexiglas wall. It looked at first as if you could reach straight ahead and grab the banana. But if you did, you hit the almost invisible barrier; to reach the fruit, you had to bend your hand around it. At first neither the chimpanzee nor the orangutan could figure it out. Both pushed their hands forcefully toward the fruit. After hitting the barrier twenty times, the orangutan paused and studied the experimental arrangement. On her next attempt, she reached around the barrier and snagged the banana. The chimpanzees tested on this device never solved the problem.

"Chimpanzees are strong, and they're used to using their force to get things," Lonsdorf said. "They're also easily distracted by what's going on socially in their group. If one of the females is in estrus, they really don't seem able to concentrate, and that affects their performance on our tests. They want to see what *she's* doing, and which males are interested in her. But if a female in Kwan's group is ready to breed, he'll take care of that and then get right back to his test."

The age, sex, social position, and personality of the apes all affected how well individuals did on these cognitive tasks, Ross and Lonsdorf said, which made it more difficult for scientists to generalize about the capabilities of the Chimpanzee Mind or the Gorilla Mind.

"We always want to generalize, to say *the* chimpanzee can do *x* or *y*. But there's a tremendous amount of individual variation," Lonsdorf said. "I think that's something we haven't figured out how to deal with yet."

As for how the chimpanzees and gorillas remembered the symbols and sequences, Ross suspected that they use something like a photographic memory, since they lack words. Perhaps, as they looked around the world, they were taking mental images, rather like snapshots, something that animal scientist Temple Grandin has suggested for animals in general. One chimpanzee in Japan so excelled at sequence-memory tests that he had become a sensation on YouTube, Ross told me. The prodigy chimpanzee, Ayumu, lives with his mother, Ai, at the University of Kyoto's Primate Research Institute, headed by Professor Tetsuro Matsuzawa. Like Lonsdorf and Ross, Matsuzawa studies both wild and captive chimpanzees and is a pioneer in using touch-screen tests to explore their minds. Because conditions in captivity are different from those in nature, scientists increasingly combine both wild and lab studies, especially if they're interested in exploring how a species evolved specific cognitive skills.

"Professor Matsuzawa's work is what really inspired our program here," Ross said. "You ought to meet him, and Ai and Ayumu," the star touch-screen chimpanzees. "The Japanese researchers have a very different relationship with their chimpanzees than we do. It's much more equitable, and it's led us to change many of the ways we handle our apes."

Later that afternoon, I met up with Ross in his office to say good-bye. I had just watched the other troop of seven chimpanzees, led by Hank, fishing for ketchup from their fiberglass termite mound—a spirited competition in which the chimps stood elbow to elbow while dipping their sticks into the mound's termite-sized openings. Years ago, I'd seen chimpanzees at Gombe hunting termites in much the same way. They'd foraged for their stick tools and chewed the end to make just the right kind of implement for the job. So it was at the zoo. Ross nodded as I described all that had gone on at the artificial termite mound—there had been a small squabble among the females; the young male, Optimus, whom Hank saw as a possible contender, had pretended to not want to join the others at the mound; Chuckie, a young female in estrus, had used her sexual receptivity to win a good fishing spot next to Hank.

"Watching the chimpanzees at that termite mound really tells a lot about the chimpanzee mind," Ross said. "You can see it so clearly there, how important their social relationships are to them—even more important than food sometimes. A happy chimpanzee is one that is living in a complex society with his fellows."

And so it would be with the boys in blue denim. Ross said their owner had at last agreed "to do the right thing." The two chimpanzees would soon be settling into the Little Rock Zoo. The general arrangements were in place, but Ross had a few details to work out, and I left him to his task.

On my way out, I spotted Keo sitting at his window. I nodded my head and waved good-bye, and was rewarded with a slight nod in return—just enough of a recognition to make me smile.

Over his life span, Keo had experienced nearly the full gamut of our ever-changing human view of what it means to be a chimpanzee—a trained clown; an "it" rather than a "he"; a mindless zoo exhibit; and these days, a human relative, one worthy of being treated as an emotional, sentient being in need of cognitive enrichment. He had not been used as a medical research animal, or as a space chimpanzee (as the chimpanzees in NASA's space program were called), or as Tarzan's sidekick in Hollywood. For that, Keo had been lucky.

Now Keo lived down the hall from Ross. They couldn't see each other from their rooms, but Ross often kept his office door open, and he could hear Keo and the others calling and hooting, their voices sometimes reaching high-pitched crescendos or dropping into lower, breathy pants that meant they were laughing. During our discussions, I noticed that Ross, like a parent, listened with half an ear to the cries. He liked hearing the chimpanzees' noises, even if some shouts meant a quarrel was under way. These were the sounds of chimpanzees living together as they were meant to be—in a social group with all of its complexities, as I had seen them at Gombe.

THAT EVENING in my hotel room, I looked at the articles and video links Ross had given me about Professor Matsuzawa's work with the chimpanzees at Kyoto University's Primate Research Institute. These were startling, not just because the young male, Ayumu, was a memory whiz, but because of the relationship between the chimpanzees and the researchers. Here were photos of a chimpanzee and scientist sitting side by side on a bench, no glass wall, not even a metal screen or bars to separate them. The chimpanzee was working on a puzzle task while the researcher took notes. In one of the videos, two white-coated scientists speak softly and encouragingly to a female chimpanzee who has given birth but has not picked up her infant. The mother is clearly distressed and frightened, her lips pulled back in fear. One of the scientists tells her not to worry, to be brave, it is your baby, and you must nurse it. A researcher gently picks up the baby and places it in the mother's arms. The infant's hands latch on to her fur. The mother then attempts to nurse but cries when the baby takes her nipple; she seems about to drop her infant to the floor. But then the soft voice of the scientist is heard again. Yes, yes, he says soothingly, it may hurt at first, but soon it will not. And slowly, the mother settles down, cradling her baby against her breast and letting the infant nurse.

Throughout the video, the scientists talked to the chimpanzee mother as if she were fully capable of understanding, almost as if they were talking to a young, human mother. They appeared to treat her as an equal and she responded in kind. Did the scientists actually feel this way, or was it merely an illusion because they were inside the enclosure with the chimpanzees? And, if they did, how had they managed to put aside the "us" versus "them" attitude that has colored so much research into the chimpanzee—or any animal—mind?

I e-mailed Matsuzawa at once, and a few months later I was on my way to Japan.

ALTHOUGH AFFILIATED WITH the University of Kyoto, the Primate Research Institute, which Matsuzawa directs, lies about ninety miles to the east in the city of Inuyama. I reached it via bullet train. Aside from the trains speeding through the city, nothing seemed to move very fast in Inuyama. Set along the shores of the Kiso River, it felt and looked like a pretty country village. The primate institute sits on a gentle rise above the town, its five-story, cement-slab building backed up against the coniferous forest of Dog Mountain. Off to one side, secluded behind fences and trees, is the outdoor chimpanzee enclosure. Here, several metal towers, one fifty feet high, provide the chimpanzees the best views in Inuyama—and a spot for broadcasting their hoots of bravado.

Only minutes after Matsuzawa greeted me in his office on an early January morning in 2009, our meeting was interrupted by a shouting chimpanzee. Matsuzawa, who was then fifty-nine and had the groomed, conservative style of a Japanese businessman, turned to look out his window, which faced the apes' outdoor arena. He was wearing a dark gray suit and striped tie as if about to greet a meeting of corporate executives, not a community of chimpanzees.

But as a new round of hoots erupted, Matsuzawa rose from his chair. "Please, come outside," he said to me, stepping onto his balcony.

"That is Akira, our dominant male," Matsuzawa said, looking toward the tallest tower.

On the tower's platform, Akira stood upright, his black fur glistening in the morning light, his fellow chimpanzees and all of Inuyama at his feet. For a moment, he looked like an actor in a Hollywood production, *Son of King Kong*. He stomped about, then let forth with another long, whooping cry.

When Akira finished, Matsuzawa nodded slightly, almost formally in acknowledgment. He then stepped to the balcony railing, opened his mouth wide, threw back his head, and let forth his own string of hoots.

"Hoo-hoo-ooo-ooo-Whoo-hoo-Whooooo—WHOOOO!"

Akira listened, not interrupting, waiting for Matsuzawa to finish, before he called in return. Back and forth they called to each other. Matsuzawa let Akira have the last hoot and then turned to lead the way back inside his office.

"Now all the chimpanzees know Matsuzawa is here," he said. He smiled as he said it, but he wasn't joking. That noisy exchange between man and ape powerfully reinforced what I'd seen in the video—in this lab, Matsuzawa and his fellow researchers are very much a part of the chimpanzee community.

Only two days before, Matsuzawa had returned from a field trip to Bossou, Guinea, in West Africa, where a colleague, Yukimaru Sugiyama, had started a long-term study of wild chimpanzees in 1976. Matsuzawa took over from Sugiyama in the late 1980s and now makes annual monthlong research visits. At Bossou, Matsuzawa began his pioneering field experiments, setting up tests for wild chimpanzees that replicate his tests in captivity. "In the lab, you can get ideas about how chimpanzees' minds work," he explained about his dual projects. "And in the forest, you can test your ideas against wild living chimpanzees."

He's used his method to investigate how young chimpanzees learn their cultural traditions and how these traditions spread to other populations. Those studies, among many others, have earned him a Jane Goodall Global Leadership Award, and a Purple Ribbon for outstanding academic achievement from the Japanese government.

Given his success, I was surprised when Matsuzawa told me in his lightly accented English that "in fact, no" he had never planned or wanted as a child to study chimpanzees. "That was never a dream of mine," he said. "I wanted to be a philosopher." At Kyoto University, he began his philosophy studies but then discovered the field of psychological physics. "I realized that it was a better field to answer my questions," he said. Soon he was implanting electrodes in rats' brains and recording the animals' responses as a way of understanding what had come to interest him the most: how the brain interprets what the eyes see. After earning his doctorate in 1976, he landed a research position at the primate institute. He switched to monkeys to continue his brain research but no longer used invasive methods.

Then, one day, Matsuzawa's advisor told him that the institute would be receiving a young female chimpanzee—its first ape. His advisor asked Matsuzawa to design a language experiment for the chimpanzee.

"I knew nothing about chimpanzees," Matsuzawa said, his eyes widening at the memory of this assignment. "I had done only experiments with rats and monkeys."

A few days later, he met the chimpanzee. He named her Ai, Japanese for "Love."

"Ai was in the basement," Matsuzawa recalled. "It was all concrete, with no windows, and a single lightbulb hanging from the ceiling. In the middle of the room, there was a log, and a tiny, tiny chimpanzee, thirty centimeters high [about 12 inches] sitting on it. I had no idea about chimpanzees, and I was so amazed. She looked into my eyes, and I looked into hers. If she had been a monkey, she would have screamed. But no. She just sat quietly and gently, looking. So I knew this is not a monkey, this is something new to me, a different creature."

Unsure of what to do, Matsuzawa made a gesture that won Ai's trust. He unfastened the cuff on his white lab coat and gave it to her. She held the cuff for a moment, then slipped it onto her own arm and ran it up and down her limb like a bracelet. Then she slipped the cuff off and handed it back to the young professor. A monkey, Matsuzawa said, probably would have eaten the cuff or torn it apart.

"So, right from day one, Ai began to teach me something new," Matsuzawa said. "I realized she is a special creature, and I took her out of the basement and put her in a cage on the sunny side of the building. It was not such a good life for her here at first, but we have learned, and made many improvements."

Matsuzawa's first meeting with Ai triggered a question that he has spent his career trying to answer: How does a chimpanzee see and understand the world? The encounter led him to develop his first computer touch-screen test, a simple exam to find out if Ai could match colors and shapes—which, in turn, led to a more sophisticated setup and experiments to see if chimpanzees can learn numbers (yes); whether they can recognize the faces and vocalizations of their fellow chimpanzees (yes); whether they perceive optical illusions (yes); whether they understand what type of tool a fellow chimpanzee needs to solve a particular problem (yes); and how young chimpanzees learn from their mothers.

Matsuzawa glanced at his watch. It was 8:40 a.m., time for us to head to the lab, he said, since the tests began at nine o'clock sharp—the same time that chimpanzees would set out on their morning travels if they lived in a forest. During his college years, Matsuzawa had been a member of the Academic Alpine Club of Kyoto, climbing peaks over twenty-five thousand feet high in the Himalayan and Karakoram ranges; he'd even been part of a Japanese team that climbed the third-highest mountain in the world, Nepal's Mount Kangchenjunga, although he did not summit. So from his office, we walked at a mountaineer's brisk pace through the institute's long hallways, and up and down several flights of stairs to the lab annex, arriving promptly at 8:55 a.m.

Fourteen chimpanzees, nearly all captive bred and spanning three generations, form the institute's colony, which closely approximates the social structure of a wild chimpanzee community. The chimpanzees live in cages inside the annex building, but as at Lincoln Park Zoo they can range into the fenced enclosure to climb ropes and the tower or to find private places in the abundant shrubbery. When it's time for "work," as Matsuzawa called the daily testing sessions, the researchers summon the apes, who travel into the building through a network of overhead catwalks. As we entered the annex, a couple of chimpanzees scampered above us, hooting excitedly.

"It is important always for the chimpanzee to be higher than we are,"

Matsuzawa said. "In the wild, they would be in trees above us, looking down. We like to give them that feeling here, so they can feel relaxed."

Inside the testing room, two large Plexiglas booths served to separate humans from chimpanzees. On the human side were tables stacked with computer monitors, televisions, video cameras, and bowls of chopped fruits. (Matsuzawa's students prepare all the chimps' food and treats, and clean up after them, too, part of the training that helps them "really understand what it is to be a chimpanzee," he said.) There were also a dozen chairs, but sized for children, not adults—although adult scientists were perched on them. The tiny chairs kept the humans low to the ground, while the chimpanzees could sit in the higher position on benches inside the booths. They reached their computer touch screens through slots cut into the Plexiglas walls.

We took our seats, our knees pressed toward our chins, and awaited the arrival of Ai and her ten-year-old son, Ayumu. His father was Akira, the group's dominant male.

"Don't worry when Ayumu comes in," Matsuzawa said. "He may spit at you and hit the glass. If you don't react, he will stop. But if you show you are afraid, he will behave even more badly. He can be a naughty boy."

Ai and Ayumu would be working on their tests separately in the two booths—much as human mother-child pairs are tested by psychologists. We could hear the two chimpanzees hooting and hurrying our way. At exactly 9:00 a.m., one of the researchers pushed a button, a door in the testing booth clanged open, and Ai and Ayumu entered. Ai was thirty-three years old, her black fur thinned from overgrooming by other female chimpanzees, her facial skin blotched and slightly grizzled from age. When she saw Matsuzawa, whom she'd known for thirty-two of her thirty-three years, she made a soft pant-hoot, then sat in front of her touch-screen monitor. It was attached to an exterior machine that would dispense food rewards whenever she solved a problem correctly.

Ayumu, whose name means "Walk," sauntered into his booth, acting saucy and frisky. He spotted me—the stranger—at once. He pressed up against the glass and spit toward me, but it wasn't a very threatening act and was easy to ignore. Instead, I kept my eyes on his mother, who had given me the slightest of nods—just enough of one to let me know she accepted my presence. Her calmness and my own behavior—staying quiet and expressionless, as Matsuzawa advised—soon had the desired effect, and Ayumu stopped his antics. Meanwhile, Ai pressed a white button on her screen, and the test flashed up.

"Ai is doing a matching test," Matsuzawa explained. "She will see a Japanese symbol for a color, and she must then press the correct color." To find out if Ai fully understood the meaning of the symbol, the test became progressively more difficult. For instance, the symbol for red would sometimes be rendered in blue or another color, but Ai still needed to press the red square. When humans are given a similar test, we automatically slow down to think it through. The researchers were timing Ai's responses to see if she showed a similar delay as she patiently worked her way through colors and symbols, receiving a bit of carrot or apple for each correct answer.

While Ai worked at her computer, Ayumu plopped in front of his, where he tackled various memory tests like the ones I'd seen him doing on YouTube. On the screen, the numerals 1 through 9 flashed up, each one the same size, and colored white. But they were scattered randomly across the screen. When Ayumu touched the first numeral on the screen, solid white blocks replaced all of the numbers, obscuring them. Ayumu was expected to remember the location of the numerals, and touch the white blocks in the correct ascending order. In other words, he had to memorize the numbers and their positions and retain that memory—and he had to do so very quickly, because the numbers were covered by the blocks within 650, 430, or 210 milliseconds, depending on the test. At 210 milliseconds, the numbers barely flashed onto the screen before being masked by the boxes. It was impossible to track each number's position with one's eyes; you had to take in the entire pattern with a single glance. Yet Ayumu nearly always got the sequence right. His success rate was close to 80 percent.

"It means that he has an actual picture memory, an eidetic memory," Matsuzawa said. "He takes a picture with his mind and holds it."

Ayumu can hold that image for a long time, too. Later, back in his office, Matsuzawa showed me a video of Ayumu in the middle of a test. Something crashes in the lab, and the chimpanzee turns away from the screen to see what it is. Incredibly, when he returns to the test ten seconds later, he casually picks up where he left off—and continues to touch the boxes in the correct numerical order.

"You or I, we cannot do this," retain a short-term memory for that length of time, Matsuzawa said. "It is something special for the chimpanzee mind. It is not a matter of training for them. It is their way of seeing the world."

That afternoon, I tried one of the basic memory tests. Nine numbers flashed on the screen for the full 650 milliseconds (to at least give me a chance) and then were replaced by the white blocks. I could get four—and once even five—in the right order, but only with great concentration. And as the test speeded up, my success plummeted. I was no match for Ayumu.

Ayumu had learned about the computer tests while sitting on his mother's lap as an infant. His mother was good at the test, but Ayumu was always better. Whether Ayumu will retain his ability as he ages remains to be seen. "It's something we would like to know," Matsuzawa said. For chimpanzees, who lack a complex language, an instantaneous flash memory would likely be very useful, Matsuzawa added, especially in social situations, where another individual's expression can quickly change.

Ayumu, as an infant, never struggled or tried to touch the computer screen while sitting on his mother's lap. He just watched. In time, Matsuzawa gave the young chimp a small touch screen of his own, which Ayumu cautiously explored. When Matsuzawa began studying how young wild chimpanzees at the Bossou site in Guinea learn to use stones to crack open oil palm nuts, he noticed the same behavior. The mother chimpanzees did not actively teach their infants. They did not take their child's hands and press their fingers around the stones, as a human mother would do. They simply let their children watch.

"It is the master-apprentice method of teaching," Matsuzawa said. "The mother does not offer any explanation. And her child, the apprentice, learns by careful observation."

The mother does not mind if her youngster steals the kernel from the nut she has just cracked open. She never scolds, or hits, or acts impatient. She places another nut on her anvil stone and hits it with her hammer rock, while her toddler stands close by, watching, and snitching the tasty kernels. In time, Matsuzawa said, the young chimpanzee will play with the stones or try to put the nut on the stone; it learns by observation and trial and error. By age four or five, when the youngster is finally old enough to grasp the stone, he or she will finally begin to successfully crack open nuts on his or her own.*

Despite the lack of overt teaching, the chimpanzees are learning their cultures from their mothers and others in their communities, Matsuzawa and others say. "They learn the local way of doing things," he explained.

The majority of primatologists accept the idea that chimpanzees—and many other species of primates and other animals, such as whales and dolphins—have cultures. That is, different populations within a species have their own way of doing things, and these ways are not genetically inherited but learned.

Cultures can spread from one group to another also by learning, Matsuzawa said. Once he and his team observed a young female from a neighboring community move into a new group at Bossou. In her old community, the chimpanzees fed on Coula nuts by placing them on a stone anvil and cracking them open with a rock hammer. In her new community, the chimpanzees broke only oil palm nuts in this way. As an experiment, Matsuzawa's team set out some Coula nuts for the female's new group. All of the chimpanzees—except her—either ignored the nuts or tried, unsuccessfully, to bite through the skin. But she didn't hesitate. She cracked open the Coula nuts on her anvil stone and ate the kernels. Soon a group of juveniles gathered to watch. They crowded around, and, about a week later, two of the youngsters were successfully copying her behavior.

Matsuzawa says the experiment explains how chimpanzee cultural tradi-

* Christophe Boesch, who studies the chimpanzees in the Ivory Coast's Tai National Park, has reported two cases of mother chimpanzees actively teaching their offspring. The mothers repositioned the nuts so that they were in a better position for the youngsters to hit with their hammer stones; and one mother showed her daughter how to hit the nut. But these are the only known cases—a very small sample given all the hours Boesch, Matsuzawa, and other researchers have watched mother chimpanzees cracking nuts with their children. Passive teaching is the norm in chimpanzee society.

tions can spread in the wild. Others are not entirely convinced, but he argues that female teaching, even if only by example, offers the best explanation. "The females spread the traditions because females are the ones who move into new communities; they are the ones who take their culture with them and introduce it to others."

LATER THAT MORNING, after the testing session ended, Matsuzawa excused himself and stepped outside the lab to slip into blue coveralls. Then, standing tall and erasing all emotion from his face, he entered Ai's booth. "Sit!" he ordered the chimpanzee, who easily could have torn Matsuzawa limb from limb. Ai quickly complied, while making soft pants of delight. "Be good," Matsuzawa said. They sat down on the floor together, facing each other. Ai proceeded to unbutton Matsuzawa's shirt, the equivalent, he later told me, of her grooming him. She liked opening the buttons on his cuffs—perhaps because it reminded her of their first meeting, when he had unbuttoned his cuff and handed it to her. He, in turn, picked at her fur, and examined her teeth. They pawed at and wrestled with each other, Ai making excited calls, Matsuzawa mimicking her wide-mouthed expressions. They played in this manner for a good ten minutes, and then he stepped into Ayumu's booth and did the same.

It was an extraordinary display of love and trust—and only possible, I thought, because Matsuzawa also understood the importance of sitting at times in the kindergartner's chair.

BEFORE I LEFT, Matsuzawa pulled a scientific journal from a pile on his desk. "You must see this," he said, holding it aloft. "There is a paper in here that is very surprising to me. In it, the authors say that humans 'with practice' are as good as chimpanzees at our memory test."

Matsuzawa frowned in exasperation.

"Really, I cannot believe this. With Ayumu, as you saw, we discovered that chimpanzees are better than humans at one type of memory test. It is something a chimpanzee can do immediately, and it is one thing—*one* thing—that they are better at than humans. I know this has upset people; you can read their comments on YouTube, where they even say it must be a trick—the chimpanzee is cheating. And now here are researchers who have practiced to become 'as good as' a chimpanzee!

"I really do not understand this need for us always to be superior in all domains. Or to be so separate, so unique from every other animal," he said. "We are not. We are not plants; we are members of the animal kingdom."

Matsuzawa placed the journal on his desk, and closed his eyes for a moment. When he opened them, he switched subjects and began to tell me about his first trip to Bossou, which he'd made in 1986 when he was thirty-five. He thought it very likely that this was the region where Ai had come from, and he wanted to make a pilgrimage there in her honor.

"I wanted to visit her hometown, her home forest," Matsuzawa said, "just to see where she was born. I am one hundred percent sure that her mother was shot when the hunters captured her. Probably her father, too, and maybe others who wanted to protect her."

On that trip, Matsuzawa hoped for nothing more than a brief glimpse of chimpanzees living in the wild on their own terms. He knew that spotting wild animals is not always easy, so he decided to spend two months in the field with his colleague Yukimaru Sugiyama; surely in that time he would see a chimpanzee. On his first morning at the camp, he and Sugiyama climbed a nearby hill, and Matsuzawa got his wish. Spellbound, he watched as a mother and her toddler traveled through the treetops, the youngster clinging to her fur, the mother deftly swinging from limb to limb and vine to vine.

"The sun was shining on their black hair as they passed over my head and disappeared into the forest. So beautiful! I thought, maybe, this could have been Ai and her mother." Matsuzawa paused for a moment. "I don't have the right words to speak about the emotional aspect of the chimpanzee," he continued. "But from that trip I learned the wild chimpanzees are the heroes and heroines of the forest. It is the opposite of here in the lab. There, they are the masters, and we are the students.

"Some people, even some scientists who study them, really do not like chimpanzees," he continued. "But I do. I think of them like a different kind of human, ones who have long, flowing black hair and who live in the forest. Really," he said, "what is so threatening to us about the mind of the chimpanzee?"

10

OF DOGS AND WOLVES

Humans created dogs in a cheerful spirit

and in their own likeness.

VILMOS CSÁNYI

One day in late November 1989, while setting out on a hike with his wife, Eve, in Hungary's Kékes Mountains, Vilmos Csányi (pronounced Chai-nee) spotted a stray dog. The dog—small, fuzzy-furred, and male—proceeded to follow them. After about five miles, the dog seemed to tire, so Csányi picked him up and carried him the rest of the way. The next day, Csányi, Eve, and the dog—whom they had named Flip—returned to the Csányis' home in Budapest—a day earlier than they had originally planned. The inn where they were staying wouldn't allow dogs in the room—and now they had one. Or the dog had its humans. Either way, the unexpected friendship also unexpectedly opened the door on a new field of research: the study of the minds of dogs.

At the time that he and Flip found each other, Csányi was the head of the Ethology Department at Eötvös Loránd University in Budapest. He and his students were focused on the learning abilities and mechanisms of the paradise fish and had made several major discoveries. Their research was well known and highly regarded, and they had an excellent international reputation.

"But this fish research was becoming very boring," Csányi told me one evening in February 2009 over tea in his book-lined study in his Budapest flat, with his elderly blond, mixed-breed dog, Jerry, at his feet. Flip had died some years before, and Csányi was now silver-haired and retired. "We were repeating our fish experiments with mice, and we were publishing good papers, but I felt we needed to change our focus."

After bringing Flip home, Csányi had started a diary about him. (He included Jerry after acquiring him eight years later.) He noted that Flip appeared to pay close attention to what his humans were saying and to understand what was about to happen, such as a walk, a car ride, or the arrival of food. He made notes about things like how Flip would make him realize there wasn't any water in Flip's bowl; or how Flip would ask him questions by sitting in front of him and looking at him "questioningly"; or how Flip would use his nose to poke Csányi in the morning to rouse his master from bed. He also began to share his dog stories with the members of his lab.

"Yes! Csányi would come in and tell us some unbelievable story about Flip, and ask us, 'Why do you suppose he did that?'" recalled Ádám Miklósi, one of Csányi's former students. "And then he would say, 'You boys figure out an experiment to test that, to show why Flip can do that.' Everyone would have a very long face. Because in those days, no one would think that you are a serious scientist if you are studying dogs."

Dogs weren't always viewed so negatively by animal cognition scientists. Darwin peppered his books on evolution with examples of dogs to illustrate his concepts and turned to them repeatedly in his study of the universality of expressions and emotion in humans and animals. The pioneering behavioral scientists Ivan Pavlov, William T. James, and D. O. Hebb also found dogs to be fine subjects for their studies; while John Paul Scott and John Fuller investigated the links between dogs' genetics and their social behaviors in a long-running project that they summarized in their classic 1965 book, *Genetics and the Social Behavior of the Dog*. Yet by the 1970s, dogs were being shunned largely because cognitive scientists and ethologists decided that domesticated animals were artificial and not as intelligent as their wild brethren. As supporting evidence, scientists pointed out that dogs' brains are about 25 percent smaller than those of wolves, their immediate ancestor. Presumably, dogs lost some degree of intelligence in exchange for living a softer, less challenging life with us.* Researchers also thought that the bond between human and dog made it impossible to study them objectively. Those few who braved going against the tide were snubbed. "My colleagues said, 'Why don't you study *real* animals?'" Marc Bekoff, a cognitive ethologist and the author of several books about animal cognition, remembers being asked when he first proposed studying dogs' play behavior in the 1970s. These days, Bekoff's studies are widely cited.

Csányi never worried about whether his lab would be derided for studying the minds and behaviors of dogs. "You would never make a scientific discovery if you thought only of your colleagues' reactions," he said. Besides, he was as intrigued as Darwin had been by the transition from wolf to dog, and because so few people were studying dogs Csányi thought canines offered a rare research opportunity. "Of course, I saw the many dark expressions [on my students' faces] when I would tell them some story about what Flip had done that weekend," he said. "Okay, that wasn't science. But these kinds of things, my stories, eventually gave us ideas for experiments—good experiments."

Experiments, it should be said, that changed the science of animal cognition and ethology, helping to expand both fields. Today researchers embrace the study not only of dogs but of domesticated animals in general, which are no longer considered simply dumbed-down versions of their feral ancestors. Dogs, in fact, are now regarded as excellent subjects for understanding how evolution can shape and transform a species' brain and for studying the mental building blocks that involve social cognition. Research programs and laboratories devoted to dog cognition have sprung up at universities from Australia to Japan, and there are even special dog cognition conferences.†

---

* Researchers still don't have an explanation for the difference between the sizes of dogs' and wolves' brains—or for why every species of domesticated animal, from ducks and geese to horses and pigs, also has a smaller brain than its wild ancestor. The reason or reasons domestication always leads to smaller brains are hotly debated; but the effects are universal. Anthropologists have documented the same change in *Homo sapiens*: the brains of modern humans have shrunk about 10 percent over the last ten thousand years.

† Another sign that animal cognition scientists have fully embraced dogs as a research subject comes from a simple count of the number of dog and canid papers presented at the Comparative Cognition Society's annual international conference. In 1994, at the group's first meeting, not one paper was about dogs. At their seventeenth meeting, in 2010, more than 10 percent of all the presentations were about dogs and their relatives. There has yet to be a similar boom in cat cognition studies.

Csányi's students, though, could not foresee these changes. All they knew was their professor had an idea that he wouldn't stop talking about—that in order to live with humans, dogs had to give up some of their wolfish ways and develop new social skills. They had to become loyal to their new group—humans—and accepting of our rules. Most of all, Csányi argued, they had to become adept at reading and responding to human communication cues. In short, they had to become mentally more like us. Csányi then proposed a bold and overarching hypothesis to guide the lab's studies: that the transformation of the wolf, *Canis lupus*, to the dog, *Canis familiaris*, was a better model for understanding the evolution of the human mind than the transformation of chimpanzee to human.

"Csányi's main hypothesis was that the minds of dogs, because they were shaped by humans, resembled the minds of their creators or inventors, so we could learn a lot about the human mind by studying those of dogs," Miklósi said. "It was a big idea—a great idea—but of course we didn't see it that way."

When Csányi broke the news in 1994 that the lab would be switching to dogs, "the great idea" struck Miklósi "actually as a bad one, in fact, a terrible one," he said, recalling his despair. "I was just at the beginning of my career, and I thought for sure this would be the end of it. I was sure no one would take us seriously. I remember thinking, 'My God, are you crazy?' That's what I thought, although, of course, I didn't say it. And I wasn't sure what we were going to do; *how* were we going to study dogs?"

THE IDEA THAT DOGS are more like us than they are like wolves goes back at least to Darwin, who wrote about their transformation, "Dogs may have lost in cunning . . . yet they have progressed in certain moral qualities, such as affection, trust, worthiness, temper, and probably general intelligence." Darwin, who loved dogs and owned many, regarded them as one of the best examples of the power of both natural and artificial selection—an animal that through its relationship with us had gained some emotional and cognitive similarities to humans, yet retained the physical qualities of its predatory ancestor. They were his prime example of how animals, other than humans, also experience "pleasure and pain, happiness and memory." Dogs possess imaginations, Darwin believed, and the power to reason, to feel love, jealousy, pride, and something akin to a conscience and religion. He reached his conclusions simply by observing his own dogs and paying attention to the reports of other dog owners—all anecdotal findings in today's world, and not acceptable to modern science journals. "Nonetheless," as Erica Feuerbacher and C. D. L. Wynne noted in an article about the history of dog cognition studies, "his ideas have often proved correct."

Csányi's students, of course, would have to use more rigorous methods than Darwin's astute but largely anecdotal observations. For Miklósi that was one of two problems about the challenge of studying dogs. No scientific standards or methodologies were then in place for studying dog cognition. It was one thing to raise and study fish in a laboratory, but how would Csányi's students obtain enough dogs for research projects that required at least fifteen subject animals for a proper statistical analysis? The second problem was more personal. Miklósi was not a dog lover. He had never had a dog—either as a child or as a university student. He does not own one today, even though after Csányi retired Miklósi moved into his mentor's position, becoming head of the Ethology Department and director of the Family Dog Lab. All his life, Miklósi, who is also the author of *Dog Behaviour, Evolution, and Cognition*, has never had a dog of his own. Miklósi loves cats.

Miklósi, who was in his midforties when we met, reminisced about his initial despair over Csányi's plan while we talked in his office in the university's biological sciences building, a simple cement-slab structure overlooking the Danube River. Outside, it was a wintry day, the ground frozen, the sun shining weakly on the river. Inside, it was comfortable enough to keep a tropical freshwater turtle alive. Miklósi had one on his credenza, housed in a shallow aquarium.

"Yes, I know it amuses everyone that I'm not a dog person," he said when I smiled. "I'm not one of those people whose eyes go all soft and crinkly whenever they see a dog. My daughter is. So we do have a dog, a sheltie, at home. But Scottie is her dog."

Miklósi had given me a quick tour of the Family Dog Lab when I first arrived. It was easy to tell it was a dog lab because there were dog owners in the hallways, and dog-owning professors, and dogs in all the offices that we passed—until we got to Miklósi's. The only animal (aside from us) in his tidy, carpeted room was the turtle. Perhaps, I speculated aloud, he had a turtle because with all the dogs around the lab, it didn't make sense to keep a cat.

"No, the turtle is my son's," Miklósi said. Miklósi had brought it to his office because he thought the turtle—whose name is Pillow—was being neglected. Here, with all the people coming and going, Pillow had a more stimulating environment than in his son's bedroom, where he was often just a lonely turtle stuck on a shelf.

Miklósi has a lean build, a mustache and small goatee, and straight, light brown hair that falls over his ears and forehead and that he sometimes has to brush away from his eyes. His fingers are long and slender, his speech is fast (even in English), and his way of moving is quick, almost catlike. If I'd seen Miklósi in a police lineup of biologists and ethologists, I would have picked him as the statistician or computer modeler—someone designing projects with virtual or imaginary animals, not the one working with living creatures. Yet he is actually the kind of person who worries about a turtle's state of mind. And so, despite his wariness of dogs, and his inability to "have a close bond with a dog," he'd found a way "to appreciate them," as he put it, and then to study them.

After deciding that their lab would focus on dogs, Csányi made sure there was no turning back by having all the aquariums and fish supplies removed; the only remaining signs of the lab's previous incarnation are some unused sinks and tiled backsplashes. Miklósi and his fellow postdoctoral student József Topál, who is now at the Institute for Psychology at the Hungarian Academy of Sciences, began reading the scientific literature on canine behavior, looking for any studies that might help them devise an experiment.

"We did nothing really but read for the first three years," Miklósi said. "We had nothing to publish. All of our first ideas were stupid. It took a lot of time to figure out what we wanted to do and could do, and then how to do it."

Csányi ended up selecting the first experiment—one that would test the long-standing scientific assumption that dogs, with their smaller brains, are not as smart as wolves. Very few studies comparing dogs' and wolves' cognitive abilities had ever been done; one of the most cited involved only four young wolves and four equally young malamutes. The canids were given the task of figuring out how to open a gate after watching a person do it. The wolves solved the problem immediately; the dogs never even tried, despite being shown repeatedly how to remove the latch.

"Professor Csányi really disliked this study," Miklósi said. "He said it was easy to explain because the dogs were well trained and well behaved and knew they shouldn't open the gate unless they had their owner's permission to go out. They knew there was a rule, and they were obeying the rule."

At least that was Csányi's hypothesis. To prove that the dogs were following a rule, the researchers needed to test various breeds of dogs of differing ages and genders, so that their results would apply to all dogs, not just those of a particular type. Where were they to find so many dogs? Csányi had already ruled out having a group of lab dogs kenneled at the university; that would have been expensive and unkind to the dogs. The researchers briefly considered using canines from a shelter but discovered that the dogs competed to be chosen for an experiment—and those that were frequently selected were later punished by the others. ("Dogs," Csányi would conclude about this behavior, "do not like exceptions." In other words, dogs have a sense of fairness.) As an ethologist, Csányi wanted to use healthy dogs that were normal and living in their natural environment—the "wild jungle" of the family home, as he called it. He also realized that he would need owners to participate because the dogs would perform best if they received directions from their masters. Finally, he turned to a dog agility club for his human and canine participants. (Csányi's method of using dogs and their owners is now followed at nearly all dog cognition labs.) Of course, every dog owner knows that his or her dog is smart, and in no time, the scientists had plenty of applicants.

Csányi's team devised a modified version of the gate-latch test. They placed cold cuts into ten dishes with long handles and slid the dishes beneath a low wire fence that ran parallel to the floor. After calling a dog and its owner into the room, one of the researchers demonstrated how he could get the food by pulling on a dish's handle. He then stepped aside, and the dogs were given three minutes to decide what to do. Some of the test subjects were outdoor dogs, used to being on their own, and so more independent than dogs that lived inside a family home. Most, but not all, of the outdoor dogs didn't hesitate; they immediately copied the demonstrator and solved the test. The few outdoor dogs that did not grab the handles were dogs that were very closely bonded with their masters. Tellingly, *all* the dogs that lived inside with their human families did nothing, aside from frequently glancing at their owners. Not one tried to touch the handle of one of the dishes. Only after their owner spoke to them did these dogs do so. Even then, some of the dogs kept looking at their owners, seeming to want their help.

"That was our first breakthrough," Miklósi said. "It showed that dogs could solve problems as wolves do but they also have this special desire to cooperate with their masters, to follow a command. They want to work with us."

People often talk about the pack nature of wolves, Miklósi added, and how wolves cooperate on the hunt and in raising their pups. So the tendency to cooperate is there in dogs' ancestors. But to cooperate with an entirely different species, one not closely related on a genetic level—and to seek that cooperation above all else—means that something has fundamentally changed in the brains of dogs.

As a dog owner, I wasn't surprised to hear the results of these tests; I doubt if most dog owners would be. What did surprise me was learning that Csányi and his team had published this study in 1997: in the late twentieth century, science had finally rediscovered the mind of the dog.

The Hungarians were soon on a roll. In 1998, the team published two more landmark papers in respected and high-profile science journals. One investigated the idea that the relationship between dogs and humans is a "social attachment"—a term that psychologists first used in the 1970s to describe the strong bond between mother and child. While other researchers had suggested that dogs behave as if they had a social attachment to humans, the Hungarians were the first to confirm this via their experiments. They revealed, for instance, that dogs that are strongly bonded to their owners experience separation anxiety when their masters leave them, just as young children who are strongly bonded to their mothers do when their mothers step away. Tightly bonded toddlers and dogs also respond similarly when mothers or owners return, racing forward to greet them and leaping and cavorting with joy. "The bonding processes of dogs and humans are truly extraordinarily similar," Miklósi said, "and they last throughout both species' lifetimes."

The second paper demonstrated that dogs pay close attention to what humans are pointing to or looking at—something that chimpanzees have great difficulty doing. "We showed that a dog can read and use a person's social cues," Miklósi said, even though dogs don't have fingers for pointing and never make a pointing gesture with their legs (although some breeds, such as pointers, do point with their entire bodies).

In this test, which is now regarded as a classic, the scientists asked a dog's owner to hide food in one of two scent-proof containers. The owner then looked at or pointed to the container with the food. It was up to the dog to use this cue to understand that his owner was in essence telling him that he knew where the food was hidden and that the dog would find it if the dog followed his owner's finger or eyes to the right location. No other animal had been shown to have this ability, so Miklósi was braced for attacks when the study was published. Instead, another team of scientists in the United States made the same discovery independently, and the two papers appeared almost simultaneously.*

"It meant that we weren't alone in doing this research; there were other scientists who were thinking along these same lines," Miklósi said. "That was when everything began to change. Of course, some people were skeptical about dogs being able to do this [to follow human pointing cues], but because there were two papers saying the same thing, it was difficult to argue against us."

By the time of my visit, Miklósi, Topál, and their colleagues had published more than seventy papers on the minds of dogs. From their studies, we now know that dogs long to be with humans almost from the moment they

---

* In 1994, Brian Hare, then an undergraduate at Emory University, stumbled into studying dog cognition after his mentor, Michael Tomasello, raised the question of whether chimpanzees understand what another chimp—or human—is thinking. Hare responded that he didn't think the skill could be that difficult because, after all, "my dog does that." Tomasello dared him to "prove it," and the result was a study almost identical to that of the Hungarians on dogs' abilities to understand the cues people give them by pointing or looking at something. Hare carried out his experiment with his family dogs in his parents' garage. He is now a leading researcher in dog and primate cognition at Duke University.

open their eyes. In tests at the lab, four-month-old puppies given a choice between going to a human companion and going to another dog preferred the person. Young wolves that members of the lab had hand-raised showed no such preference. The researchers discovered that even adult dogs living in rescue shelters rapidly form attachments to people; it took a mere thirty minutes of interaction between a person and an adult shelter dog for the dog to begin forming a bond. It's rare in most species—other than humans—for adults to form attachments, and to the Hungarians, this was another indication that dogs possess humanlike traits.

In another test of dogs' social learning capabilities, the scientists showed that dogs can imitate a person's action, much like children playing Simon Says. The volunteers would stand in front of their dogs and turn in a circle, or jump up, or place an object in a container, while calling out enthusiastically in Hungarian, "Do as I do!" Most of the dogs readily performed the action, succeeding as well as sixteen-month-old children who were given the same test.

The Hungarians revealed, too, that dogs have a sense of what's "right" and "wrong" and can follow our human rules, a social skill that helps to strengthen the bonds within a group. Dogs even understand that human rules aren't necessarily rigid; they can be flexible—and humans inconsistent, sometimes telling them that it's okay to get on the couch and other times demanding that they get down.

Perhaps most surprisingly, with their colleagues at the Hungarian Academy of Sciences, the scientists showed that dogs will follow human cues on certain tasks despite evidence right in front of their eyes that they are making the wrong decision. In this test, a researcher repeatedly placed a ball in Box A, while making eye contact with the dog and explaining that he was hiding the ball there: "Rex! Watch, Rex. I'm hiding the ball here." Then Rex would be allowed to find the ball. In the next phase, with Rex watching, the experimenter silently moved the ball to Box B. Where would Rex now search for the ball? In Box A. Human toddlers make exactly the same error because both dogs and young children learn by paying attention to the social cues adult humans give them, rather than by making their own judgments—one of the most striking examples of how similar the minds of dogs have become to those of humans.*

As a result of the Family Dog Lab's experiments and tests at other dog cognition labs, dogs are now hailed as a natural experiment, one we humans set in motion when we brought dogs into our homes and lives some fifty thousand or more years ago. Many researchers now agree with Csányi's initial belief that dogs and humans represent an unusual case of what is termed convergent evolution, because although dogs and humans have entirely different ancestors, we share numerous behaviors and traits—particularly the desire to work together to accomplish a task.

"It has been a complete shift," Miklósi said. "I'm as surprised as anyone, because really, I didn't think we were going to make any discoveries. And from what we've discovered, we can say some general things now about dogs. I may not be a dog lover, but even I can see the special, dynamic relationship dogs have with us. With cats, we have a very simple social relationship; it's not so much about doing things together. But that is really at the heart of the dog-human relationship. They are companions, yes, but they are also with us in working relationships, like herding, guarding, or going to war. We do these things together with them. Cats hunt *for* us, but we don't hunt *with* them. It's a big difference."

Miklósi glanced at his watch. The student of one of his colleagues was about to start an experiment. We walked down the hallway to the office of Péter Pongrácz, the student's supervisor. Miklósi tapped on the door and cracked it open a few inches.

"It's okay," Pongrácz said. "My dogs will stay."

Miklósi and I agreed to meet later, and I slipped inside Pongrácz's office, where four dogs lay curled up next to his feet. They lifted their heads and ears, studying me. One was chestnut colored, the others were completely black; all were medium-sized, with long, pointed noses and curly body fur, although the fur on their faces was smooth.

* There are differences between a dog's and child's responses to this test. If a different person moves the ball to Box B, the dog will follow his eyes and go to Box B. Switching adults doesn't affect a child's decision; he or she will still search for the ball at Box A, because toddlers are essentially programmed to receive instructions from all adults.

"These are my dogs," Pongrácz said. "They are *Mudis*, Hungarian shepherds, and they are very smart."

Pongrácz was tall and beefy, with a long, oval face and dark blond curly beard, which made him look rather like a blond *Mudi*. His dogs watched us briefly, then lay back down.

Were they going to take part in the test? I asked.

"No. We use them to try out our ideas for experiments. But then for the real test, we want dogs that have never seen the experiment, so that we can show they have not been trained."

Pongrácz was investigating why dogs bark, what they might be communicating. Wolves bark, too, but only to warn or to protest. Miklósi had told me that dogs "invented barking as a means of communicating to us as much to other dogs" and that they "can modulate the frequency or pulse" to signal fear or feeling lonely or playful. One of Pongrácz's previous studies showed that humans can easily identify the differing barks of a lonely, fighting, or playing dog. "That means that a dog's bark is often directed at us to convey the dog's inner state," Pongrácz said. "Maybe they do this because they live with a very talkative species."

Pongrácz's student, Tamás Faragó, would be giving the test, which was designed to further investigate the communication between dogs and people. He and Pongrácz spoke briefly in Hungarian, and then Faragó said in somewhat halting English, "Please, join us."

In the hallway outside, a middle-aged woman had a bright-eyed Cairn terrier on a leash. "This is Kopé," Faragó said, introducing me first to the dog and then to Kopé's owner. Faragó directed us into the lab room, where he had arranged a small cage that was covered with a cloth. Inside the cage was a concealed tape recorder, and outside lay a large, tempting bone.

Faragó stood next to me in the back of the room and directed the owner to let Kopé off his leash. Immediately, Kopé spied the bone and made a beeline for it. But just as Kopé reached the bone, Faragó used a remote trigger to start the recorder. From the cage came the deep-throated growl of another dog, and Kopé froze in place.

"That's a food-guarding growl," Faragó whispered to me. "As soon as a dog hears that, he knows he better leave that bone alone."

When the unseen dog growled a second time, Kopé, looking unnerved, ran halfway back to his owner, wagging his tail. He peered up at her face and then looked back over his shoulder at the bone.

"He's asking for help," Faragó explained. "'Come on, Mommy, help me get that bone. Let's do it together.'"

And that, Miklósi said when we met up again that afternoon, is what is fundamental to dogs' cognitive abilities: their strong desire to work with and for us, and their ability to communicate even without language—with nothing more than a soulful look.

"We've joined forces with the dog," Miklósi said. "We've brought their minds into our lives, and into our work as companions and assistants, as our collaborators. Looking back now, at when we first started, I have to say I'm surprised that no one wanted to study dogs before, or thought dogs were mindless. These days, we understand dogs in a completely new way. It's like having new glasses on."

WITHOUT THE DISCOVERIES of Csányi and his students, scientists might not have paid attention when a border collie named Rico appeared on the popular German television show *Wetten das?* in 2001. He knew the names of some two hundred toys, and to the delight of the television audience he enthusiastically fetched them by name from his collection. Julia Fischer, then a postdoctoral fellow at the Max Planck Institute for Evolutionary Anthropology in Leipzig, Germany, happened to be among the millions of viewers. She told her fellow student Juliane Kaminski about Rico, and the two arranged a meeting to test the collie. They discovered that Rico possessed an uncanny language ability: he could learn and remember words as quickly as a toddler, they reported in *Science*. When the scientists placed a new toy among his familiar toys and gave it a new name, Rico managed by using some type of simple logic ("I know the others, but I've never seen this one, so it must be it") to almost always retrieve the new toy. Even after not seeing the toy for a month, Rico could successfully fetch the toy in half of his trials. Without repetitive instruction, he had integrated the toy's name into his vocabulary.

Other scientists had shown that two-year-old children—who acquire around ten new words a day—have an innate set of principles that govern how they learn and remember these. The capacity for recalling new words is one of the key building blocks in language acquisition. Fischer and Kaminski suspected that the same principles guided Rico's word learning and that the technique he used for learning and remembering words—termed "fast mapping"—was somewhat similar to that of humans.*

To find more examples, Fischer and Kaminski read all the letters from hundreds of people claiming that their dogs (and in some cases, cats) had Rico's talent. In fact, only two—both border collies—had comparable skills. One of them—whose owner asked me to call her dog by the pseudonym Betsy to protect her from "dog haters"—has a vocabulary of more than three hundred words.†

"Even our closest relatives, the great apes, can't do what Betsy can do—hear a word only once or twice and know that the acoustic pattern stands for something," Kaminski told me during a visit with her and her colleague Sebastian Tempelmann to Betsy's home in Vienna in 2006. They were giving Betsy a fresh battery of tests. Kaminski petted Betsy, while Tempelmann set up a video camera.

"Dogs' understanding of human forms of communication is something new that has evolved," Kaminski explained, "something that's developed in them because of their long association with humans." Although Kaminski has not yet tested wolves, she doubts they have this aptitude. Because the skill has been found only in border collies, she thought that it was somehow related to their traditional herding jobs—that over the years shepherds had selected those dogs that were the most highly motivated and attentive listeners.

How similar are the border collies' language skills to those of humans?

---

* Inspired by Rico, a retired psychologist in South Carolina, John W. Pilley, has taught his border collie, Chaser, 1,022 words. Chaser could have learned more, Pilley told a *New York Times* reporter in 2011, but the human was bored.

† I interviewed Kaminski and Betsy's owners for my *National Geographic* article on animal minds; the owners were proud to have their dog featured, but they worried that someone who disliked dogs would recognize her from her photograph in the magazine and harm her.

For abstract thinking, we employ symbols, such as words, letting one thing stand for another. Is this what the dogs were doing, too? That's what Kaminski and Tempelmann hoped to discover through their tests with Betsy.

Betsy's owner, who asked to be identified by the pseudonym Schaefer, summoned Betsy. She obediently stretched out at Schaefer's feet, eyes fixed on her face. Whenever Schaefer spoke, Betsy tipped her ears forward and attentively cocked her head from side to side.

Kaminski handed Schaefer a stack of color photographs and asked her to choose one. Each image depicted a dog's toy—such as a teddy bear, a mini-Frisbee, or a plush toy lobster—against a white background. Betsy had never seen these toys, and of course they weren't actual toys; they were only images of toys. Could Betsy connect a two-dimensional picture to a three-dimensional object?

Schaefer held up a picture of the rainbow-colored mini-Frisbee and gave it a name, "Frisbee." "Look, Betsy," she said in German, as if speaking to a young child. "This is Frisbee. *Frisbee.*" She said the name several times, tapping the picture and urging Betsy to look. Then she placed the photo face down behind her and told Betsy, "Now, go find Frisbee. Find Frisbee."

Betsy ran into the kitchen, where the Frisbee was placed among three other toys and photographs of each toy. She came running back triumphantly with the Frisbee photograph, and after Schaefer dispatched her again, brought back the Frisbee itself.

"It wouldn't have been wrong if she just brought the photograph," Kaminski said. "But I think Betsy can use a picture, without a name, to find an object. It will take many more tests to prove this, and even then, there will be critics because this is a kind of abstract thinking."*

In his observations about dogs, Darwin noted that they had probably evolved something he termed "general intelligence" by becoming domesticated. Dogs had a mental flexibility he thought wolves did not possess. They also have very breed-specific talents, such as the border collies' language

* Kaminski now suspects that dogs "may perceive a lot of human communication as commands to do something." Although we humans may tell our dog to "find the ball" or "find the Frisbee," the dog may regard the name alone as an order. Thus a dog may think the sounds "ball" or "Frisbee" also mean "Find the ball" or "Fetch the Frisbee."

abilities, or the now extinct "turnspit dog" that Darwin mentioned, whose task was to run inside a treadmill attached to the meat-roasting spit on the family hearth. Dogs' willingness to help us in so many varied ways is perhaps the most remarkable thing about them—and certainly a key distinguishing feature between them and their wolf ancestors.

"Some people say, 'Oh, well, the dog is just another domesticated animal,'" Miklósi had said to me toward the end of my visit at the Family Dog Lab. "But it is not. Dogs were the *first* domesticated animal, and not just by one thousand years, but by ten thousand or twenty thousand years, maybe longer. That is a huge advantage. And then they were taken very rapidly by people everywhere, I think via trading. You can imagine how everyone would want this precious new, little thing."

The first known dogs were not, in fact, little, although in time we developed smaller breeds. Although scientists argue about when dogs were first domesticated, recent discoveries suggest that the earliest dogs were large, powerfully built animals.* In 2012, paleontologists working at Předmostí, a site in the Czech Republic that dates to about twenty-seven thousand years ago, reported the discovery of the remains of three dogs from an encampment where people once hunted mammoths. Judging from the skeletal material, these dogs would have weighed between seventy-five and ninety pounds, and stood about two feet tall at the shoulder—not as big as Eurasian wolves, but certainly the size of today's German shepherds. Intriguingly, three of the skulls were perforated—someone had punched holes in them to remove the brains—and one held a piece of mammoth bone in its mouth. The scientists think that someone—most likely one of the mammoth hunters—had placed

* Studies comparing dog and wolf DNA indicate that dogs separated from wolves about one hundred thousand years ago, about the same time that our species, *Homo sapiens*, traveled from Africa into Southeast Asia and Europe. So it may be, as Csányi has suggested, that modern humans began domesticating the wolf not long after first meeting them. However, some archaeologists insist that these early wolf-dogs or incipient dogs were not dogs as we know them; they argue that truly domesticated dogs do not appear until much later in the archaeological record, when there is solid proof of the human-dog bond. For these scientists, the earliest domesticated dog dates only to fourteen thousand years ago—the date of a burial site in Germany where a dog and human were found interred together.

the bone in the dog's mouth as part of a ritual to "feed" the soul of the dead animal. Perhaps the hunters had also removed the dogs' brains to release their souls, or more practically, to use for tanning hides. Even older dog remains, dating to thirty-two thousand years ago, have been uncovered in Belgium, while the thirty-three-thousand-year-old skull of an "incipient dog" was reported from a site in Siberia's Altai Mountains in 2011. These, too, were large animals, yet clearly distinguished from wolves by their shorter and broader snouts and crowded teeth.

What would these early dogs have been like? How much do dogs actually differ from wolves in their minds and behaviors?

While the Hungarians had taken a stab at this question in 2001, Miklósi urged me to visit another group of scientists who were hand-raising wolves in Austria to see the difference for myself.

WHEN I FIRST MET ZSÓFIA VIRÁNYI at the Wolf Science Center in Ernstbrunn, Austria, she looked as weary and rumpled as anyone would who'd spent the night sleeping—or trying to—with six four-month-old wolf pups in a barn. Bits of straw and lint clung to her clothes, her eyes were puffy and rimmed with dark circles, and her curly, honey-blond hair was not yet combed. As she walked to the center's gate to greet me, she pulled a piece of grass from her jacket and dusted off her pants. "Well, this is how it is with wolf puppies," she said. She spoke with the voice of experience, since, while earning her doctorate in ethology, she'd worked with Miklósi and his colleagues at the Family Dog Lab on their study of hand-raised wolves. Indeed, Virányi knew wolves almost better than she knew dogs. Not only had she lived with, bottle-fed, and slept with a wolf puppy during that first experiment, she'd subsequently helped raise a dozen more wolf pups for this new project; she'd also studied captive wolf facilities throughout Europe and the United States while researching the best methods for managing wolves in captivity.

Virányi's German colleague Friederike Range, who'd driven me from Vienna to the center, nodded in sympathy as Virányi rubbed her eyes. Even though Range had not spent the night sleeping with the wolves (they were old enough now that the wolf raisers could alternate nights), she also looked exhausted—the result, she'd told me on our drive, of the four months' nonstop work of hand-raising the pups.

"You never get a full night's sleep," Virányi said. "Six hours, maybe. They're very restless in the early night and early morning. For them, it's time to hunt. And in the middle of the day, when you want to do something with them, they only want to sleep. Their natural rhythms are very different from ours—and from dogs."

Range and I had arrived just after 8 a.m., so the pups were still wide awake. They'd been sitting with Virányi and Tódor, her black-and-tan, mixed-breed dog, inside a fenced enclosure at the privately owned Ernstbrunn Game Park about thirty miles north of Vienna. At the time of my visit in September 2009, the Wolf Science Center's headquarters were housed here in two old stone buildings that had once served as the stables for an aristocrat's estate; today, they're located in brand-new buildings also inside the park. In addition to the wolf pups, three adult wolves that the scientists had also hand-raised lived in another outdoor arena, amid the park's rolling oak woodlands. Deer, wild boar, and wild sheep freely roam the grounds, but the wolves are strictly confined. The three oldest wolves had come to the center from a zoo in Austria; two of the pups were brought from a zoo in Basel, Switzerland; and the other four pups came from a game park in Montana. All had been separated from their mothers and littermates before they were ten days old so that they would not develop a wariness of humans.* Their round-the-clock care and feeding had been taken over by Virányi, Range, Kurt Kotrschal, a zoologist at the University of Vienna, and a crew of volunteers as part of a remarkable experiment to investigate the relationship between wolf, dog, and human.

"We want to understand this triad," Range had explained on our drive, her English rising and falling with the soft inflections of her German accent. "People say many things about wolves, dogs, and humans—how they're alike or how they're different. But the truth is we're only at the beginning of un-

* Other researchers discovered that wolves need to be socialized very early in their development if they are to accept humans. Wolf pups that aren't removed from their litters until they are eight to ten weeks old develop an extreme wariness of humans and try to avoid them.

derstanding the minds of wolves and dogs, and understanding how humans changed the wolf to make the dog. In fact, we're at such a basic level we have almost everything to learn."

By hand-raising wolves and a separate group of dogs, the scientists could track, test, and compare the cognitive development of both animals—something that had never been done in such a controlled manner, or with dogs and wolves that had comparable upbringings from their pup stage into adulthood. (As of July 2012, the scientists had hand-raised eight more wolf pups and thirteen dog pups in the same way.) And with those results in hand, they could then see if dogs through domestication had truly become more humanlike in their minds and behaviors.

At the center, Range, with her border collie, Guinness, in tow, had led the way to the wolf pups' enclosure to introduce me to Virányi. As soon as they spotted us, the pups yelped with excitement, an adult wolf howled in response, and Tódor and Guinness barked. Virányi unlocked the enclosure's gate to let herself and Tódor out, but she shooed away the pups and carefully refastened the lock, then checked it again, before turning to shake my hand.

The wolf pups—one black, one blond, and four mostly tan and gray—stood on the opposite side, whining and pawing at Range and Guinness. The pups noticed me, too, but gave me only brief, wary glances. If I looked their way, they dropped their heads and peered up at me with their shoulders hunched, bodies angled, and legs askew—the pose of a cautious but curious animal.

"Yes, there's a stranger here," Virányi said to them. "Someone new to meet this morning." To me, she said, "New people; new things—they always get a little bit anxious."

"They'll warm up to you," Range said. "Just move slowly; let them come to you." She knelt close to the fence and called the wolves' names. "Hello, Nanuk and Yukon. Why didn't you let Zsófia sleep last night? The black is Apache; the blond is Cherokee. And those," she said, pointing at the last two, "are Geronimo and Tatonga." The scientists used "Indian names," they said, because their pups—even the two from the Swiss zoo—were North American gray wolves, and the names were meant to honor their heritage.

"It's good for them to meet new people," Range said. "It helps them learn how to form relationships with people, which is what we want. We aren't trying to be part of their pack, but we want them to respect us and to follow our rules in certain situations. But we also respect them, and so follow their rules in other situations." For instance, when inside the wolves' enclosure, the scientists consider themselves as the guests of the young wolf pack and are careful to never compete with them over objects or food; they also don't take sides in any of the pups' disputes but let them sort out their contests over their dominance hierarchy. "This way we can ask the wolves to do things, more or less like you would ask a dog."

As we talked, I watched the pups. Perhaps our ancestors had brought young wolves like these into their camps. If you just glanced at them from a distance, you might think they were German shepherd puppies. They were about the same size and build, with legs that seemed too long for their bodies. They nipped and growled at each other, and tumbled about as puppies do. But there was something else about them that I couldn't identify; some element that wasn't at all like a dog puppy.

"We'll go in to meet the pups now," Range said. The wolves would know if I was fearful, she added, and that would make them even more anxious, because they would not understand that I was afraid of them. "Just act normal," she advised.

Virányi said, "Zip up your jacket. Make sure your shirt is tucked in. They may get attracted to anything that's loose or hanging from your pockets. They would like to pull it."

I nodded. It wasn't hard to imagine why a wolf would like to pull something that was hanging from a body: *What are those? Entrails?* They were naturally curious, and a dangling shirttail would prove irresistible. Some years before, I'd met the wolves at another research facility, Wolf Park, in Battle Ground, Indiana. "Kneel down on their level to greet them," the scientist there had instructed me. "They'll want to welcome you to the pack. They'll put their teeth on your head. It'll be okay." I knelt, and the wolves crowded in close, yipping with excitement, and brushing up against me. First one, then another, and finally a third took turns fastening their teeth lightly around the top of my head. From the corner of my eye, I could see that they needed to stand up briefly on two legs to do this. They opened their jaws wide to encompass my skull but didn't bite; it was just a quick graze of fangs against skin. Perhaps because I've always loved dogs, I wasn't frightened and didn't pull away, but I've never forgotten the sensation, either.

Virányi unlocked the gate, and one by one we all went inside. She laughed and called to the wolves as if she were calling dog puppies. The wolves yipped and ran after her and Range, while Guinness and Tódor, who work with the wolves alongside the scientists, trotted close to their humans. The dogs held their heads up stiffly as if they were police dogs on assignment; they didn't make any playful moves toward the pups. I followed, joining Virányi and Range beneath an apple tree. We sat cross-legged on the ground while dogs and wolves sorted out their own positions. Tódor, Virányi's dog, took a seat square in the middle of her lap and barked sharply at the pups, while Guinness ran circles around us, keeping the pups corralled. The wolves had to wriggle and worm their way past Guinness to get to us.

"Do the pups think of the dogs as their elders?" I asked.

"Yes," said Range. "Their *mean* elders. They let the pups know that they are dominant."

"Actually, the dogs are a little bit afraid of the wolves," Virányi said. "They were even afraid of them when the pups were just a few weeks old. They know that these are not dogs, even if they're puppies."

Two of the pups managed to crawl onto Range's lap, while a third lay down next to Virányi. The scientists petted and fussed over them as one does with puppies. Tatonga, one of the gray-and-tan wolves, hesitantly approached me. She whimpered, wagged her tail, and inched closer when I spoke to her, imitating the way Virányi and Range talked to the pups, in a high, happy voice. "It's okay. You can come close." And she did, working up her courage to lick my hands and chin, and even giving me a gentle and friendly nip there. At last, somewhat relaxed, she tucked in her tail and placed her head in my lap so that I could stroke her. In time, all the pups came to greet me, licking and squirming around me as Tatonga did, and I petted them all. But I don't want to give the wrong impression. I never felt that these wolf pups were completely at ease with me, or I with them. It wasn't like sitting with a group of playful, loving dog puppies. Maybe the tension was partly due to Guinness and Tódor, who remained ever alert, lunging and barking sharply at the pups to keep them in line.

"That's how the pups learn to behave," Range said. "They've learned to pay attention to us and to follow some commands—to sit, lie down, stay—by watching the dogs. They do everything the dogs do"—which helped during the scientists' behavioral experiments, when they asked the dogs to demonstrate what they wanted the wolves to do.

Virányi and Range were both in their thirties, and their style and personalities—Virányi, warm and outgoing, Range, more serious and cerebral—complemented each other. Their manner of relating to the wolves and dogs, on the other hand, was nearly identical: they were relaxed, patient, and affectionate—an ease that suggested they'd spent a good part of their lives in the company of animals, as was the case. Between them they had studied sooty mangabeys, capuchin and squirrel monkeys, marmosets, ravens, keas (a parrot species that lives in New Zealand), pigeons, gorillas, and orangutans. But dogs and wolves were now their passion. "It's now my life's work, forever, I think," Virányi later told me.

Virányi had brought a long leash with her, and she clicked it onto Tatonga's collar so that we could take the wolf for a walk and give her a social learning test. Every day, the wolf pups were led on these individual walks and given various tests to assess their skills as they grew up. From the center, we walked down a narrow lane to a meadow edged with conifers and oaks. Along the way, three students who worked at the center walked toward us; Tatonga knew them all, Virányi said, yet the pup tucked her tail between her legs and slowed her pace.

"Say something!" Range shouted to them. "Call to Tatonga!"

"Tatonga! Tatonga!" they shouted back. Tatonga pricked her ears forward, wagged her tail, and ran forward to greet them.

"She knows these girls very well," Range said, as we continued on our walk. "You'd think she could just use her nose to smell them. We always hear so much about how dogs and wolves have such great noses. And then you see them do something like this. Sometimes even Guinness seems not to recognize me from a distance; then when I get close to her, or call her name, she acts exactly like Tatonga did, like she's saying, 'Oh, it's *you*.' Why doesn't she just pick up my scent?"

The canine sense of smell was yet another example, Range said, of a trait we ascribe to dogs and wolves but about which we actually know very little. It was on the scientists' lengthy list of canid abilities they wanted to investigate.

When we reached the meadow, Virányi first did a quick training session to teach Tatonga to look her in the eye, something that is difficult for wolves, as it is for most wild animals, because staring is generally a threat. She asked Tatonga to sit a few feet in front of her and held up a treat for her at arm level. Each time Tatonga met her gaze, Virányi snapped a clicker (a simple training device) while simultaneously tossing her the treat. Sometimes Tatonga looked straight at Virányi, other times she tilted her head from side to side, as if unsure where to focus. "She can only manage it for a second, or she'll look at me from the corner of her eye," Virányi said after the two-minute lesson. "The wolves know they must do *something* to get that treat, but they really aren't sure what it is. They think maybe it's about moving their head in a particular way, that's why she does that funny head-wobble. And even when they figure it out, they can't do it like a dog does, looking openly into your eyes. When I speak to Tódor, he turns instantly to focus on me, to gaze at my eyes. It's like he's asking, 'What do you want to do?' Or, 'What do you want me for?' Wolves don't think that way; they're too busy thinking for themselves."

Next, the scientists gave Tatonga a social learning test that they give the wolves once a month. "We want to know at what age wolves can begin to learn something about their environment from humans, or from their conspecifics," meaning their fellow wolves, Range explained. She emphasized that "nothing is known about social learning, or learning overall, in wolves; none of these tests [which comparative psychologists developed and use regularly with other animals from rats to pigeons] have been done before with wolves."

The test was simple. First, away from Tatonga's line of sight, Virányi walked on three different paths, each one the same length. When she returned, Range shortened Tatonga's leash, while Virányi held the carcass of a baby chick just above her head, letting her sniff at it. The wolves are fed largely during their training exercises—"They must earn their food," Range had said earlier—and Tatonga was hungry. She tried to leap at the chick, but Range restrained her from grabbing it. Virányi then turned and walked about twenty feet away along one of her original paths. At the end of this trail, she dropped the chick near some shrubs, and then, walking along the same path, returned to Tatonga. She showed her empty hands to the pup, turning them palm up for the pup to smell. Range kept her on the leash but let it play out so that Tatonga could go where she wanted. To my surprise, the wolf simply stayed put, never so much as taking a step on any of Virányi's trails. Yet the pup had watched the human—who held the dead chick—walk to a very specific place. I thought about my own dog, Buck. Even as a four-month-old pup he would have figured out that I must have left the chick for him to find; he would have gone in search, and if he couldn't find it on his own, he would have looked to me for help. Tatonga did none of these things. She sniffed the grass around her, gazed at the forest, and finally lay down. After two minutes, the test was over.

"Stupid, stupid, stupid wolf!" Range said, laughing, since nearly all the other wolf pups had easily passed the test. "You lost your chick." In the wild, though, this type of failure would have serious consequences, since young wolves that did not pay attention to their caretakers' or parents' actions would likely not survive.

Tatonga didn't seem to mind; it was as if she'd forgotten the bird entirely after Virányi displayed her empty hands. "The wolf puppies are really not like dog puppies; they don't have that interest in everything we do," Virányi said as we walked Tatonga back to the pups' enclosure. "They are a lot more analytical and have some strong ideas, goals, and interests of their own. And they're never really relaxed like a dog or a puppy. They're almost overly sensitive; they jump away from you when you think that there's no reason. And they're always alert, just as you saw with Tatonga when we met the students. It's as if they're expecting danger, or that they're more alert to the idea that things are uncertain."

The first time Virányi raised a wolf pup, she was living at her parents' home. Virányi arranged a sleeping area for herself and the pup in the living room, but the pup grew restless at night, and no matter what Virányi did to soothe her, she wanted to roam. "We put up a plywood board between the two rooms, and she got it in her mind that she wanted to get over it. She just would slam against it, for an hour at a time." Then the pup would sleep for an hour, but as soon as she woke up, she started in again. "No matter how many times I corrected her, or told her no, or led her to another place with food rewards, she went right back to slamming against that board. She would just throw herself at it, over and over. Very different from a dog."

As another example, Virányi described how differently the young wolves and dogs reacted when they first touched the electrical fence that surrounds the wolves' enclosure. She had touched it herself to understand the buzzing sensation. "It didn't really hurt, but it is a weird feeling." When one young wolf received her first shock, she silently turned back to it and, from only an inch away, closely inspected the wire with her nose. "She seemed to know this thing caused the weird feeling." In contrast, when Tódor was shocked, he ran screaming straight to Virányi and hid between her feet. "He growled in the direction of where this bad thing had happened, but he had no clue where the pain came from. If anything, he thought the people standing close to that spot had done it to him. Wolves," she concluded, "seem to take many more things into account than do dogs." For dogs, humans are "more important than anything else."

Virányi and Range had tested the first three wolves they raised at the center for their ability to understand that they should go to the spot where a person points a finger. Another experiment had shown that both eight-week-old dog and wolf puppies can do this. Intriguingly, however, the four-month-old wolf pups in Virányi and Range's test failed, as did four other wolf pups of the same age that were given the exam by other scientists at a wolf park in Hungary. "They were struggling and biting. They were just too busy doing other things to pay attention," Range said. Only when they reach maturity (around two years old) are wolves able to pay sufficient attention to understand and follow a human's pointing gesture. "They are on a different developmental path from dogs," she said, perhaps because ultimately dogs "must live in our world and obey our rules. They have to learn many of the same things that children learn."

In Budapest, when I'd been at the Family Dog Lab, I'd also met Márta Gácsi, one of the other researchers who'd helped with the lab's wolf project and who'd led the study comparing wolves' and dogs' understanding of the human pointing gesture. She had shown me a video of a student trying to pet her seven-week-old wolf puppy while it was eating its dinner. As soon as her hand drifted toward the pup's nose, the young wolf's lips curled back, and he snapped viciously. "You know, sometimes you give a dog food, but sometimes you take it back," Gácsi said, "and most dogs don't mind. You can't do that with a wolf, not even a pup. They really would bite, even though we had cared for them since they were just a few days old." The differences in how dogs and wolves mature socially and developmentally are further evidence of the genetic changes that occurred in dogs as a result of being domesticated, Gácsi said.

When I mentioned Gácsi's comment about the snapping wolf puppy, Virányi and Range smiled. They'd seen this behavior, too. Range said that people often asked her if she wouldn't like to have one of the wolves live with her in her home. "I always say, 'No,' because I like to have access to my fridge. If a wolf is in my home, then he's going to control the refrigerator"—and, I thought, everything else.

So given the headstrong, competitive, wary, and independent nature of wolves—even as pups—how had we ever domesticated them?* "I think probably some wolves came close to humans, maybe to scavenge," Virányi said. "Maybe some stayed nearby, and they began to change, to become more relaxed. They had to because even in those days you could never have a wolf—a real wolf—in your home or camp. If you did, you'd have to watch your kids nonstop."

Virányi said that probably some of the camp-following wolves had evolved to become somewhat friendly toward people. It was likely that it was their pups—pups that would have been easier to handle, and more willing

* In the late 1950s, Dmitry Belyaev, a Russian geneticist, began a domestication experiment with silver foxes, selectively breeding them for the sole traits of friendliness to humans and reduced aggression. After more than forty generations, the foxes are very similar to dogs—they're attracted to humans, they wag their tails and whimper when humans walk toward them, and they lick the humans' hands. The foxes have also changed physically and now have black-and-white coats, floppy ears, and tails that curve over their backs; they also reach sexual maturity earlier, and they bark.

to be looked at in the eye—that people first brought into their camps. "Then you could just start breeding—selecting—for these traits, for things you liked in the wolf. Actually, I have a harder time to imagine the people back then—what they were like, and how they knew to do this, to create the dog. When you think about it, it seems almost unreal, like a fantasy. How did they do it?"

IN FRANCE'S OLDEST DECORATED CAVE, the Grotte Chauvet, archaeologists have traced images of animals—lions, horses, rhinoceroses, and elephants—that the artists of twenty-six thousand years ago painted on the rock walls. They've found the skulls of cave bears and measured the tracks of cave bear paw prints. They've recorded the charcoal smudges left behind by people carrying torches to paint or perhaps to simply admire the great works of art. And in one chamber, they've found the footprint of a child. There are no other human footprints with this young person, who stood about four and a half feet tall, the scientists say. But there is another nearby footprint: a dog's. Or a dog-wolf's. They know the canid was more dog than wolf because of the length of its middle toe, which is doglike. The archaeologists investigating Grotte Chauvet cannot say for certain if the child and dog were together or if they came to the cave at separate times.

Most of us, though, would say the child and dog were surely there together and that they were friends, looking out for each other, so natural does it seem—even twenty-six thousand years ago—for humans and dogs to be a pair, working as a team.

epilogue

What cannot be denied or evaded is that this science has a moral dimension. How we study animals and what we assert about their minds and behavior greatly affects how they are treated, as well as our own view of ourselves.

DALE JAMIESON

Many books, both popular and scientific, about animal minds end with a chapter that either attempts to explain or celebrates (or both) how the human mind differs from those of the other creatures on our planet. It's an understandable urge, this desire to know how we changed from our animal brethren, why we are different. In the course of my research for *Animal Wise,* several scientists wanted to know how I planned to address the question of what makes us different. "That is what these studies are all about," one said to me. "You can't avoid it." He and others seemed puzzled when I replied that, in fact, answering that question was not the point of my book. Every time someone declares that they've found *the* skill separating "us" from "them," someone else surfaces to say they've just found that ability in another species. Given the number of new discoveries scientists are making about animals' mental and emotional lives, it doesn't seem possible to answer. Also, and most important, it seems to me the wrong question to ask. Instead, given that we now know that we live in a world of sentient beings, not one of stimulus-response machines, we need to ask, how should we treat these other emotional, thinking creatures?

It's a question that people and organizations fighting for the rights of animals used in medical research, and in pharmacological and cosmetic laboratories as well as on factory farms and in universities, have raised for many years—long before the majority of cognitive scientists were willing to acknowledge the true natures of the animals they were studying. And while some positive changes have been made in the treatment of animals, the problems are far from being fully solved. As recently as 2008, horrific videos surfaced of cattle being kicked, shocked, and prodded into a slaughterhouse's "killbox" in Chino, California. That year Californians also approved a ballot initiative that requires farms to provide enclosures for their animals that are large enough for them to stand, stretch their limbs, and turn around. The United States does not yet have a national standard for the treatment of farm animals, but an increasing number of states are addressing the matter. Based on Victoria Braithwaite's research on pain in fish and Jaak Panksepp's study of laughter in rats, it seems past time to find better methods for managing these animals when they are used for our needs.

Some cosmetic companies are also seeking—and in some instances have found—alternatives to testing animals. And changes have come to medical research, with more regulations to ensure that animals are at least given decent housing and care. Others are quitting the practice of using (certain) animals altogether. In 2008, Case Western Reserve School of Medicine became the last American medical school to abandon the traditional method of teaching cardiology: operating on anesthetized dogs to examine their beating hearts, and disposing of them after the lesson. Perhaps, one day, we will reach a point where all biomedical studies will require only the use of cells and not whole animals. There are those who hope that in the future we will also be able to grow our meat in flasks or test tubes, eliminating the need altogether for animals forced to live to feed us—although, as critics of this idea have noted, test-tube meat is not a solution to the bigger problem: our poor treatment of animals overall.

Even if we resolve the difficult ethical debates about the use of animals on farms or in laboratories, we are left with that larger issue, one that is increasingly of concern as the human population continues to surge. The United Nations recently counted seven billion of us on this planet and foresees more than ten billion by the year 2100. What does this mean for the other animals? Can we find some way to share with them what is left of the wild, natural world? Most of us, living in cities, seldom see these creatures except in the pages of magazines, on television documentaries, or in movies or YouTube videos. Wild animals are rarely part of our daily lives, and it is easy to forget how much everything we humans do affects them. Yet we live in a time of mass extinctions, an epoch scientists have termed the Sixth Extinction for the extraordinarily rapid rate at which animals and plants are dying. If biologists' predictions hold and current trends continue, as many as half of Earth's species will be gone—that is, made extinct—by the end of the twenty-first century.

ONE OF THE MORE HEART-WRENCHING PARTS of my job as a correspondent for *Science* is receiving a message from a scientist about a species that is on the verge of extinction—a beetle, a bird, a rare fish, some seldom seen rodent, a dolphin. "Can you please find some way to write about this animal?" the scientist asks. "Any attention you can give may help. We need to let the world know." I once suggested to my editor that we keep a weekly or monthly tally box, announcing that such-and-such a creature has just gone extinct—its behaviors, mind, thoughts, and ways, its beauty vanished from our planet. But even among scientists, who are the bulk of *Science*'s audience, there is little interest in this news, not enough to sustain a regular Extinction box.*

We don't have to stand by, though, and let animals—and their minds—disappear. We know how to save them. In the space of fifty years, we did a masterful job bringing back nearly all the great whale species—gray whales and blues, sperm whales and humpbacks, and the others—animals we had almost hunted to extinction. We're also making efforts to stop the decline of sea turtle species around the world by raising awareness of the turtles' plight,

* At *Scientific American*, John R. Platt does an excellent job with his blog, *Extinction Countdown*, tracking news of endangered species, and species that are recovering from the brink of extinction, http://blogs.scientificamerican.com/extinction-countdown/.

guarding their nests, and enacting laws to protect them. In the United States, when shrimp trawlers were discovered in the 1980s to be killing the turtles, we devised turtle excluder devices. These help turtles, sharks, and other large fish to escape the shrimpers' nets, reducing this bycatch. Granted, sea turtles are still highly endangered or threatened—but all these actions tell me that we are concerned for the turtles' welfare; we want them to survive. We've also restored golden lion tamarins to Brazil, the North American gray wolf to parts of the western United States, and dozens of species of birds and marsupials in New Zealand.* Hundreds of similar reintroduction and conservation projects are under way around the world.

With the whales' return, we have made wonderful discoveries. Humpbacks have cultural songs, we've learned, and these spread westward from their populations in the eastern Pacific to those in Hawaii, rather like our popular songs that seem to sweep from California to New York. Gray whale mothers, we know now, are engaged in life-and-death dramas—and battles of wits—with predatory orcas. The grays are not just swimming mindlessly on their long round-trip travels from Alaska to Mexico but must be plotting their journeys, seeking ways to evade the killer whales. "I'm sure a gray whale mother who has lost her calf to a killer whale will change her route on her next migration," Lance Barrett-Leonard, one of the scientists who tracks both the hunting orcas and the migrating gray whales, told me. "She'll remember where that happened, and she'll try another tactic to make it past the orcas the next time." The mother gray whale, after all, has a mind, and she is using it. Barrett-Leonard spoke as if that were the most natural thing for an animal to do.

It was very common in the last century to manage wild animals almost as if they were vegetable crops. Even today, whale populations are referred to as "stocks," implying that they are farmed. The prevailing wildlife management philosophy has also generally held that removing older individuals from populations—whether elk, deer, cougar, bear, or wolves—will improve the species' health and vigor. Discoveries such as Cynthia Moss's and Karen

* The wolf has been removed from the U.S. Fish and Wildlife Service's list of endangered species in Montana, Idaho, and Wyoming; the three states allow wolf hunts, so the species' spread is likely to be curtailed.

McComb's about the importance of the older elephant matriarchs to the survival of their herds are beginning to challenge this notion, as are findings about how animals think and respond to human management.

For instance, wildlife biologists regularly capture and examine wild animals by darting them with anesthetic drugs. The practice of capture, anesthetizing, and handling was thought to mildly upset the animals but wasn't believed to have lasting effects. And then in 2008, Marc Cattet, a wildlife veterinarian and researcher at the University of Saskatchewan who studies grizzly bears in Alberta, reviewed his ten years of data and examined the bears' blood. He and his fellow researchers found the blood contained far higher levels of enzymes that are indicative of muscle damage than anyone had realized. If we had these same levels of enzymes, "we'd be damned sore," Cattet said. So were the bears. When handling the bears, Cattet's research team fastened radio-tracking collars around the animals' necks and mapped the bears' movements over time. He found that the stiff and sore bears moved far more slowly than the bears that had not been captured and handled. After being collared, the grizzlies traveled half as much as they usually would for up to five weeks. And then, almost as if to make up for time, they hurried ahead for the next few weeks. "They almost seemed to have a schedule," Cattet said. Like us, the bears have places to go and things to do. Roger Powell, a wildlife biologist at North Carolina State, found exactly the same results when he looked back at his twenty-two-year study of radio-collared black bears. There were no obvious signs that the animals were suffering psychological stress from being captured and handled "but it has to have a big effect, just knowing how we would respond to something like that," Cattet told me in a telephone interview. He and Powell and their colleagues expect that all handled animals must experience physical and psychological stress. To minimize any adverse effects, they have urged their fellow scientists to improve the current procedures for handling study animals, and to explore less invasive methods.

When we don't understand how animals think about their world, we make mistakes that have unintended consequences, for ourselves and for the animals. Consider the cougars of Washington state. In 1996, residents passed a state ban on hunting mountain lions with dogs. The ban alarmed rural citizens, who feared the cougars' population would explode. Instead, wildlife biologists began recording one of the highest rates of human-caused mortality since the 1960s, when the cats were still pursued by bounty hunters. At the same time, there were more reports of lion sightings than ever before—which naturally caused only more worries for the public. What had gone wrong?

Contrary to what one might think, the increased sightings did not mean that there were more lions; the sightings meant that the lions' social order was in disarray. That's what Rob Wielgus, a wildlife ecologist at the Large Carnivore Conservation Laboratory at Washington State University in Pullman, realized as he began to study the problem. In response to the concerns of the rural residents, the state's Department of Fish and Wildlife had increased the number of cougar-hunting licenses it sold, allowing anyone who bought a deer or elk hunting tag to also buy a cougar permit for an additional ten dollars. Whenever there were lion sightings, the department issued still more licenses. At one point, some 66,000 cougar-hunting tags were being sold annually—even though the state had only about 4,000 of the cats.

By 2008, Wielgus found that hunters in his study area had killed so many adult lions, especially males, there weren't *any* four-year-old males left. In a healthy mountain lion community, the senior males (those four and older) patrol and protect their territories, and keep younger males in check—strategies that help the females to quietly raise their kittens. The older males and females also generally avoid humans. But when the senior cougars are removed, chaos ensues—just as it would if all the adults in our society were suddenly killed and only teenagers were left to handle the affairs. Wielgus's research has not gone unnoticed; Washington's Department of Fish and Wildlife is no longer automatically approving additional hunting whenever there is an increase in mountain lion sightings.

Bears, we now know, are not merely wandering randomly across the landscape looking for food and mates, and cougars are not asocial loners. We were wrong about why these animals behave as they do, in part, I think, because most of us do not grant animals even the simplest form of thought, or recognize that they do things intentionally.

The greatest threat to all wild animals is the loss of habitat—the destruction of the areas and resources they need to survive. Climate change is exacerbating the pressures, forcing some species to move to higher elevations or toward the edges of their ranges. There aren't any easy or ready solutions, and I'm not pretending to offer any here. But as we wrestle with these matters, we would do well to remember that the animals our decisions affect are, like us, beings who have minds.

Old-school skeptics and naysayers ("killjoys," as one prominent philosopher calls them) may dismiss the latest findings on animal intelligence as so much sentimental, romantic anthropomorphizing. But why is it romantic to acknowledge that animals are thinking and feeling beings? Considering the weight of recent scientific evidence, I would argue that it's actually realistic to do so. By embracing this larger understanding of our fellow creatures, we may yet succeed in overcoming the great tragedy of the Sixth Extinction.

"THERE IS GRANDEUR IN THIS VIEW OF LIFE," Darwin concluded at the end of *On the Origin of Species* about the powers of natural selection. Through natural selection, he wrote, "endless forms most beautiful and wonderful have been, and are being, evolved." By "forms," Darwin meant the splendid variety of physical shapes and structures of animals—the simple, fleshy tubes of worms; the hard, jewel-like encasements of beetles; the basic four-limbs-and-one-head body plan of vertebrates, clothed in scales, feathers, fur, or a veil of body hair. No matter how different our morphology, we animals are basically alike because of our shared evolutionary past. But animal bodies are not empty forms; they are equipped with sensory cells and brains. In his later works, Darwin argued that these internal structures—and their accompanying thoughts and emotions—evolved as well. With the "endless forms" have come endlessly beautiful and wonderful minds.

It is our good fortune to be living among them. It is a tragedy to lose a single one to extinction.

What do the minds of animals tell us about ourselves? That, like us, they think and feel and experience the world. That they have moments of anger, and sorrow, and love. Their animal minds tell us that they are our kin. Now that we know this, will our relationship with them change?

Acknowledgments

In the process of researching and writing *Animal Wise*, I talked to and spent time with the many scientists, students, and animal trainers whom I've quoted in these pages. They invited me to their labs and study sites, let me watch them at work, and generously introduced me to their animals while sharing their research and stories with me. I thank them all for these most remarkable experiences. They also read my chapters about their research and kindly sent me comments and corrections. Any errors or wild speculations are not due to these researchers but are my own.

Animal Wise grew out of a story, "Minds of Their Own," which I developed with Tim Appenzeller, former executive editor of *National Geographic*, and I thank him for this most wonderful assignment and for seeing it through to publication as a cover article in the March 2008 issue of the magazine. I also thank *National Geographic* editors Oliver Payne and Lynn Addison for editing my story. My editor Elizabeth Culotta at *Science* understands my passionate interest in animal cognition, ethology, and conservation, and has given me numerous feature assignments about these subjects, as has David Grimm, my editor at *ScienceNow*. Writing for *Science* has proved to be one of the best ways to keep abreast of these lively and dynamic fields of study.

My good friend Nancy Sosnove took time from her own busy life to carefully read and comment on an early draft of *Animal Wise*. Cornell University's Harry Greene, herpetologist, author, and lifelong student of animal behavior and evolutionary biology, kindly reviewed my historical overview of the field of animal cognition research, and pointed me to Gordon Burghardt's studies of animal play. I greatly benefitted from wide-ranging discussions with Lars Chittka, a behavioral ecologist at Queen Mary, University of London; Katharine (Katy) Payne, a specialist on elephant communication and researcher in the Bioacoustics Research Program at the Laboratory of Ornithology at Cornell University; Jack Bradbury, an emeritus professor of ornithology who is also at Cornell; and Ludwig Huber, a cognitive biologist at the University of Vienna, Austria. I'm not sure whether I would have written *Animal Wise* if I had not had the good fortune to meet Gillian MacKenzie, my agent. Her enthusiasm, encouragement, and perceptive suggestions helped me at every stage of the book.

My deepest thanks to Crown Books' Rachel Klayman, who first persuaded her colleagues to accept my book proposal; John Glusman, who read the first draft and offered excellent suggestions about how I might improve it; and Vanessa Mobley, who guided and encouraged me through the writing of the final draft. I cannot thank you enough, Vanessa, for all your confidence in me and my project—and for your patience and wise, insightful comments. It has been a joy to work with you.

My husband, Michael McRae, a journalist and author himself, listened to my stories about animals and scientists and then read and deftly edited them, too. It is difficult to imagine how I would have completed this book without his love, support, sense of humor, and fine cooking. We've watched animals together around the world, from the mountains and sagebrush plains of Oregon to the savannas of East Africa, and have raised two dogs and several cats together, and know how diminished our lives would be without our animals. To Quincie, Kitty-Pie, Buttons, and Bella—thank you for the love, laughs, and happy memories. To Nini, our headstrong calico cat, and Buckaroo, our joyful and (of course) very bright collie, thanks for spending long days and nights watching me stare at my computer when there were clearly more exciting things to do outside. And also to sweet Buck, thanks for nudging my fingers from the keyboard every afternoon. You were right: it really was time to go take a walk.

Notes

Between 2006 and 2012, I interviewed most of the scientists I've quoted in these chapters, often in their offices or laboratories or at their field sites. In some cases, I've done follow-up interviews on the telephone and have also checked details or clarified issues via e-mail messages. Unless otherwise noted, the quotations in this book come from those interviews.

vii **"Surely, the most important part of an animal . . .":** Henry David Thoreau, *The Journal* [1860], in *The Writings of Henry David Thoreau, Manuscript*, vol. 31, February 18, 1860, http://thoreau.library.ucsb.edu/index.html.

INTRODUCTION

1 **"Our organ of thought may be superior . . .":** Stephen Walker, *Animal Thought* (London: Routledge and Kegan Paul, 1983), 388.

4 **I had traveled to Goodall's study site:** The account that follows is from my visit with Goodall and her team at Gombe Stream Research Center, Tanzania, October 10–15, 1987.

10 **"[Animals] eat without pleasure . . .":** Nicolas Malebranche, quoted in Rodis-Lewis, G., ed., *Oeuvres completes* (Paris: J. Vrin, 1958–70), ii, 394; translated and cited in Peter Harrison, "Descartes on Animals," *The Philosophical Quarterly* 42, no. 167 (April 1992), 219.

10 **"Answer me, mechanist . . .":** Voltaire, *Voltaire's Philosophical Dictionary* (Fairford: Echo Library, 2010), 14.

11 **"psychology will be based on a new foundation . . .":** Charles Darwin, *On the Origin of Species* [1859], in *From So Simple a Beginning*, ed. Edward O. Wilson (New York: Norton, 2006), 759.

11 **intelligence in earthworms "has surprised me . . .":** Charles Darwin, *The Formation of Vegetable Mould Through the Action of Worms* (1881; repr., Teddington, Middlesex: Echo Library, 2007), 9.

12 **"general principles" of expression:** Charles Darwin, *The Expression of the*

Emotions in Man and Animals [1872], in *From So Simple a Beginning*, ed. Edward O. Wilson (New York: Norton, 2006), 1276.

12 **"Though led by instinct . . .":** Ibid., 331.

12 **"mind in animals":** Georges John Romanes, *Animal Intelligence* (London: Elibron Classics, 1886), vi.

12 **"we can only *infer* . . .":** Ibid., 1.

12 **"inverted anthropomorphism":** Ibid., 10.

13 **"The real question is not whether machines think . . .":** B. F. Skinner, *Contingencies of Reinforcement: A Theoretical Analysis* (Englewood Cliffs, NJ: Prentice-Hall,1969), 288.

14 **she wrote about them in ways that were scientifically forbidden:** Jane Goodall, *Through A Window: My Thirty Years with the Chimpanzees of Gombe* (Boston: Houghton Mifflin, 1990), 15.

14 **the final version had at least "conferred on the chimpanzees the dignity of their separate sexes":** Jane Goodall, *The Chimpanzees of Gombe: Patterns of Behavior* (Boston: Harvard University Press, 1986), 60.

15 **"[They] talk of all sorts. . .":** Bernard E. Rollin, "Anecdote, Anthropomorphism, and Animal Behavior," in *Anthropomorphism, Anecdotes, and Animals*, ed. Robert W. Mitchell et al. (Albany: State University of New York Press, 1997), 127.

16 **For example, Pinker says . . . :** Steven Pinker, *How the Mind Works* (New York: Norton, 2009), 24.

16 **"ferment of constructive excitement in ethology":** Donald Griffin, *The Question of Animal Awareness: Evolutionary Continuity of Mental Experience* (Los Altos, CA: William Kaufmann, 1981), x.

16 **"Good God, if mere insects communicate . . .":** Ibid., 161.

17 **"cognitive ethology":** Ibid., x.

17 **"evolutionary continuity of mental experience":** the subtitle of Griffin's book.

18 **Some scientists, such as elephant researcher Joyce Poole:** The account that follows is from my visit with Poole at her field site in Amboseli National Park, Kenya, in October 1999.

18 **ethologist Sultana Bashir spoke with sorrow:** The account that follows is from my visit with Bashir at her field site in Serengeti National Park, Tanzania, on June 8, 2004.

19n **"go through a year or two of a mourning period":** John Klavitter, quoted in Michelle Berger, "Till Death Do Us Part: Birds That Mate for Life," Audobonmagazine.org, February 10, 2012, http://magblog.audubon.org/till-death-do-us-part-birds-mate-life.

20 **thinking "tempers the raw sensory information . . .":** John Marzluff and Tony Angell, *Gifts of the Crow: How Perception, Emotion, and Thought Allow Smart Birds to Behave Like Humans* (New York: Free Press, 2012), 42.

22 **"Consciousness does not belong only to humans . . .":** Rodolfo Llinas, interviewed on "The Electric Brain," *NOVA*, October 23, 2001, transcript at www.pbs.org/wgbh/nova/body/electric-brain.html.

22 "humans are not unique . . .": Philip Low, "The Cambridge Declaration on Consciousness" (conference, University of Cambridge, Cambridge, UK, July 7, 2012), http://fcmconference.org/.

22 **"evolved, emergent qualities of brains":** Richard Dawkins, quoted in Michael Powell, "A Knack for Bashing Orthodoxy," *New York Times,* September 19, 2011.

23n **"vague, subjective descriptors . . .":** *Brain, Behavior and Evolution,* January 1992, quoted in George F. Striedter, *Principles of Brain Evolution* (Sunderland, MA: Sinauer Associates, 2005), 37.

24 **"One of the cats would do the test once . . .":** telephone interview with Immanuel Birmelin, October 4, 2011.

CHAPTER ONE: THE ANT TEACHERS

27 **"The brain of an ant . . .":** Charles Darwin, *The Descent of Man* [1871], in *From So Simple a Beginning,* ed. Edward O. Wilson (New York: Norton, 2006), 859.

27 **"Augh! What?! Augh! . . .":** James Bruffe, "The Army Ant Song: They Just Want to Breed," a rap song profile he wrote about Nigel Franks for a course at Georgetown University, February 2002.

29 **known as the "Idea Man":** Thomas Seeley's comment about Nigel Franks, quoted in Virginia Morell, "Watching as Ants Go Marching—and Deciding—One by One: A Profile of Nigel Franks," *Science* 323 (2009): 1284–85.

35 **"first non-human animals" to qualify as teachers:** Nigel R. Franks and Tom Richardson, "Teaching in Tandem-Running Ants," *Nature* 439 (2006): 153.

35 **rat mothers "do not teach their young what to eat":** Bennett Galef Jr., Elaine E. Whiskin, and Gwen Dewar, "A New Way to Study Teaching in Animals: Despite Demonstrable Benefits Rat Dams Do Not Teach Their Young What to Eat," *Animal Behaviour* 70 (2005): 91–96.

37 **"Perhaps" the flies "are little machines . . .":** Vincent G. Dethier, "Microscopic Brains," *Science* 143 (1964): 1145.

37 **"You name it and we've tried it":** Vincent Dethier, quoted in Howard Simons, "Scientist Finds Flies Can't Learn but Moths and Bats Use Sonar," *Washington Post*, April 28, 1966.

37n **Abramson regards Turner's research as "the foundation of . . .":** Charles I. Abramson, "A study in Inspiration: Charles Henry Turner (1867–1923) and the Investigation of Insect Behavior," *Annual Review of Entomology* 54 (2009), doi: 10.1146/annurev.ento.54.11807.090502.

38 **an "innate schoolmarm" in all of us:** Konrad Lorenz, "Autobiography," his speech for receiving the Nobel Prize for Medicine in 1973, www.nobelprize.org/nobel_prizes/medicine/laureates/1973/lorenz-autobio.html.

38 **When Darwin discovered that earthworms had "some degree of intelligence":** Charles Darwin, *The Formation of Vegetable Mould Through the Action of Worms* (1881; Teddington, Middlesex: Echo Library, 2007), 9, 28, 30.

38n **"Associative learning in dogs and humans . . .":** Saeed Tavazoie, quoted in Michael Hopkin, "Bacteria Can Learn," *Nature*, May 8, 2008, doi:10.1038/news.2007.360.

39n **"An individual actor A . . .":** Tim M. Caro and Marc D. Hauser, "Is There Teaching in Nonhuman Animals?," *Quarterly Review of Biology* 67 (1992): 153.

43 **one of those "special expressions of man":** Charles Darwin, *The Expression of the Emotions in Man and Animals* [1872], in *From So Simple a Beginning*, ed. Edward O. Wilson (New York: Norton, 2006), 1415.

45 **"pleas for changes in the definition of teaching . . .":** Thomas O. Richardson et al., "Teaching with Evaluation in Ants," *Current Biology* 17 (2007): 1523.

47 **"Ant colonies are no different from brains . . .":** Douglas Hofstadter, *Gödel, Escher, Bach: An Eternal Golden Braid* (New York: Basic Books, 1970), 334.

47 **"it is possible that primate brains . . .":** Thomas Seeley, *Honeybee Democracy* (Princeton: Princeton University Press, 2010), 217.

CHAPTER TWO: AMONG FISH

49 **"I wouldn't deliberately eat a grouper . . .":** Sylvia Earle quoted in Peggy Orenstein, "Champion of the Deep," *New York Times Magazine*, June 23, 1991.

49 **What is it like to be a fish?:** My opening paragraph is a variation on the question posed by Thomas Nagel in "What Is It Like to Be a Bat?," *Philosophical Review* 83 (October 1974): 435–50.

49 **a 1991 international gathering of ethologists:** The International Ethologi-

cal Congress in 1991. See Gordon M. Burghardt, "Amending Tinbergen: A Fifth Aim for Ethology," in *Anthropomorphism, Anecdotes, and Animals*, ed. Robert W. Mitchell et al. (Albany: State University of New York Press, 1997).

55 **"There was actually very little that was 'comparative' . . .":** interview with James Ha, Seattle, September 16, 2008.

58 **Morgan's Canon:** C. Lloyd Morgan, *An Introduction to Comparative Psychology*, 2nd ed. (London: W. Scott, 1894), 53.

72 **Even insects can express "anger, terror . . .":** Charles Darwin, *The Expression of the Emotions in Man and Animals* [1872], in *From So Simple a Beginning*, ed. Edward O. Wilson (New York: Norton, 2006), 1467.

CHAPTER THREE: **BIRDS WITH BRAINS**

74 **"In the very earliest times . . .":** Nalunglaq quoted in K. Rasmussen, "The Netsilik Eskimos," in *Shaking the Pumpkin*, ed. Jerome Rothenberg (Garden City, NY: Doubleday, 1972), 45.

78 **The shows "were a revelation to me":** Irene M. Pepperberg, *Alex and Me* (New York: HarperCollins, 2008), 53.

78 **she didn't stop to rethink her "new calling":** Ibid., 54.

83 **"Want a nut. *Nnn. . . uh . . . tuh*":** Ibid., 179.

87 **the brains of birds were dominated by their "basal nuclei . . .":** Alfred S. Romer and T. S. Parsons, *The Vertebrate Body*, 2nd ed. (Philadelphia: W. B. Saunders, 1977), 584.

89 **Emery dubbed them "feathered apes":** Nathan J. Emery, "Are Corvids 'Feathered Apes'?: Cognitive Evolution in Crows, Jays, Rooks and Jackdaws," in *Comparative Analysis of Minds*, ed. S. Watanabe (Tokyo: Keio University Press, 2004).

90 **"Of the species that produce . . .":** Erich Jarvis, quoted in Rebecca Morelle, "Animal World's Communication Kings," BBC News, May 1, 2007, http://news.bbc.co.uk/2/hi/science/nature/3430481.stm.

91 **The evening before he died:** This exchange is recounted in Irene Pepperberg, *Alex and Me* (New York: HarperCollins, 2008), 206.

91 **her buddy, a pal "full of life and mischief":** Ibid., 212.

91 **"Good-bye, little friend":** Ibid.

91 **"Clearly, animals know more than we think . . .":** Ibid., 219.

CHAPTER FOUR: **PARROTS IN TRANSLATION**

93 **"The animals want to communicate with man . . .":** Brave Buffalo, a Teton Sioux, quoted in F. Densomor, *Teton Sioux Music,* Bureau of American Ethnology Bulletin 93 (Washington, DC: Government Printing Office, 1918) 172.

94 **what Aldo Leopold once said he "fervently" wished for:** Aldo Leopold, "The Thick-Billed Parrot in Chihuahua," *Condor* 39 (January 1937): 9.

99 **"Man has an instinctive tendency to speak . . .":** Charles Darwin, *The Descent of Man* [1871], in *From So Simple a Beginning,* ed. Edward O. Wilson (New York: Norton, 2006), 809.

99 **"language faculty":** Noam Chomsky, *Reflections on Language* (New York: Pantheon, 1975), 12.

99 **a seminal paper:** Marc Hauser, Noam Chomsky, W. Tecumseh Fitch, "The Faculty of Language: What Is It, Who Has It, and How Did It Evolve?," *Science* 298 (2002): 1569.

100 **That means "I'm here, come to me!":** Karim Ouattara et al., "Campbell's Monkeys Concatenate Vocalizations in Context-Specific Call Sequences," *Proceedings of the National Academy of Sciences* 106 (December 22, 2009), 22026–31.

CHAPTER FIVE: **THE LAUGHTER OF RATS**

116 **"Although some still regard laughter . . .":** Jaak Panksepp, "Beyond a Joke: From Animal Laughter to Human Joy?," *Science* 1 (April 2005): 63, doi:10.1126/science.1112066.

124 **"I do not see emotions and feelings . . .":** Antonio Damasio, *Descartes' Error* (New York, Penguin Books, 1994), 164.

125n **"We see [joy] . . .":** Charles Darwin, *The Expression of the Emotions in Man and Animals* [1872], in *From So Simple a Beginning,* ed. Edward O. Wilson (New York: Norton, 2006), 1303.

127 **"Rats have the ability to reflect . . .":** Jonathon Crystal, quoted in Gisela Telis, "The Rodent Who Knew Too Much," *Science*NOW, March 8, 2007, http://news.sciencemag.org/sciencenow/2007/03/08-01.html?ref=hp.

128 **they "know what good sex is . . .":** James Pfaus, quoted in Natalie Angier, "Smart, Curious, Ticklish. Rats?," *New York Times,* July 24, 2007.

CHAPTER SIX: **ELEPHANT MEMORIES**

132 **"Who can know what goes on . . .?":** Joyce Poole, *Coming of Age with Elephants* (New York: Hyperion, 1996), 275.

149 **"They stop and become quiet and tense . . .":** Cynthia Moss, *Elephant Memories: Thirteen Years in the Life of an Elephant Family* (New York: William Morrow, 1988), 270.

149 **"It is a haunting and touching sight":** Ibid., 271.

149 **"I felt sure that he recognized it as his mother's":** Ibid.

150 **"Why would an elephant stand in silence . . . ?":** Joyce Poole, *Coming of Age with Elephants* (New York: Hyperion, 1996), 161.

155 **the cells "that make us human":** From my research, the first use of this phrase for the von Economo cells was in Helen Phillips's "The Cell That Makes Us Human," *New Scientist* 182 (June 19, 2004): 32–35.

157 **"humans of the sea":** The subtitle of John C. Lilly's book *Lilly on Dolphins* (Garden City, NY: Anchor Press/Doubleday, 1975).

CHAPTER SEVEN: **THE EDUCATED DOLPHIN**

158 **"Man had always assumed that he was more intelligent . . .":** Douglas Adams, *The Hitchhiker's Guide to the Galaxy* (New York: Dell Rey, 2009), 159.

159 **the young ape . . . appeared to be "astonished beyond measure":** Charles Darwin, quoted in Zoological Society of London Library, "Artefact of the Month," June 2008, http://www.zsl.org/about-us/library/artefact-of-the-month-june-2008,912,AR.html.

159 **"first experimental demonstration of a self-concept . . .":** Gordon G. Gallup Jr., "Chimpanzees: Self-Recognition," *Science* 167 (1970): 87.

159 **"Our data suggest that we may have found a qualitative psychological difference . . .":** Ibid.

162 **"in an apparent effort to allow [him] to breathe":** M. C. Caldwell and D. K. Caldwell, "Epimeletic (Caregiving) Behavior in Cetacea," in *Whales, Dolphins, and Porpoises*, ed. K. S. Norris (Berkeley, CA: University of California Press), 767.

162 **"There is no doubt in our minds . . .":** Ibid.

163 **"Save the seal!":** Robert L. Pittman and John Durban, "Save the Seal! Whales Act Instinctively to Save Seals," *Natural History*, November 2009, 48.

163 **"We saw a large gray whale . . .":** Dr. James D. Krueger, personal communication, February 7, 2011.

170 **"home of the world's most well-educated dolphins":** Louis M. Herman, "Facts and History about TDI and KBMML," http://www.dolphin-institute.org/about_tdi/facts_about_tdi.htm.

172 **"delphinese"**: John C. Lilly, *Lilly on Dolphins: Humans of the Sea* (Garden City, NY: Anchor Press/Doubleday, 1975), 69.

CHAPTER EIGHT: **THE WILD MINDS OF DOLPHINS**

180 **"If an alien came down . . ."**: Lori Marino quoted in Brandon Keim, "Whales Might Be as Much Like People as Apes Are," *Wired Science*, June 25, 2009, www.wired.com/wiredscience/2009/06/whalepeople/.

181 **the larger male "exhibited the greatest amount of excitement"**: A. F. McBride, "Meet Mr. Porpoise," *Natural History*, January 1940, 26–29.

181 **such an intelligent "mind in the waters"**: Joan McIntyre, *Mind in the Waters* (New York: Charles Scribners' Sons, 1974), 94.

182 **"a mind the size of a planet"**: Interview with Peter Corkeron, March 26, 2012.

196 **a chimpanzee's "feats of intelligence"**: N. K. Humphrey, "The Social Function of Intellect," in *Growing Points in Ethology*, ed. P. P. G. Bateson & R. A. Hinde (Cambridge, UK: Cambridge University Press, 1976), 307–17.

196 **"Why then . . . do the higher primates need to be as clever as they are?"**: Ibid., 307.

196 **members must become "calculating beings"**: Ibid., 309.

203 **he built a special flooded house:** John C. Lilly, *Lilly on Dolphins: Humans of the Sea* (Garden City, NY: Anchor Press/Doubleday, 1975).

204 **"humanoid" speech:** Ibid., 69.

204 ***"He does not go away,"***: Ibid., 164.

204 **Peter "devised a subtle, gentle method"**: Ibid., 177.

204 **he worked at teaching her to "trust him"**: Ibid., 182.

204 **Howe says she was "flattered"**: Ibid., 179.

CHAPTER NINE: **WHAT IT MEANS TO BE A CHIMPANZEE**

206 **"Chimpanzees bridge the gap . . ."**: Jane Goodall, *Through a Window* (Boston: Mariner Books, 2001), 249.

214 **Goodall "couldn't even talk about the chimpanzee mind . . ."**: Jane Goodall quoted in Jon Cohen, *Almost Chimpanzee: Searching for What Makes Us Human, in Rainforests, Labs, Sanctuaries, and Zoos* (New York: Henry Holt, 2010), 157.

215 **she watched her favorite chimpanzee . . . "carefully push a long grass stem . . ."**: Jane Goodall, *In the Shadow of Man* (Boston: Houghton Mifflin, 1971), 35.

215 **"the first recorded example of a wild animal . . ."**: Ibid., 37.

215 ***Homo* as a creature capable of making tools "to a set and regular pattern"**:

L. S. B. Leakey, press conference, Washington, DC, April 15, 1964, quoted in Virginia Morell's *Ancestral Passions: The Leakey Family and the Quest for Humankind's Beginnings* (New York: Touchstone, 1995), 236.

216 **they "manifest intelligent behaviour . . .":** Wolfgang Kohler, *The Mentality of Apes*, trans. Ella Winter (London: K. Paul, Trench, Trubner, 1925), 265.

CHAPTER TEN: **OF DOGS AND WOLVES**

234 **"Humans created dogs . . .":** Vilmos Csányi, *If Dogs Could Talk: Exploring the Canine Mind*, trans. Richard E. Quandt (New York: North Point Press, 2005), 272.

235 **looking at him "questioningly":** Ibid., 140.

237 **"Dogs may have lost in cunning . . .":** Charles Darwin, *The Descent of Man* [1871], in *From So Simple a Beginning*, ed. Edward O. Wilson (New York: Norton, 2006), 806.

238 **"pleasure and pain, happiness and memory":** Ibid., 800.

238 **"Nonetheless . . . his ideas have often proved correct":** Erica N. Feuerbacher and C. D. L. Wynne, "A History of Dogs as Subjects in North American Experimental Research," *Comparative Cognition and Behavior* 6 (2011): 49–71.

240 **"Dogs . . . do not like exceptions":** Vilmos Csányi, *If Dogs Could Talk: Exploring the Canine Mind*, trans. Richard E. Quandt (New York: North Point Press, 2005), 5.

240 **the "wild jungle" of the family home:** Ibid., 4.

242n **a study almost identical to that of the Hungarians:** B. Hare, J. Call, and M. Tomasello, "Domestic Dogs (*Canis familiaris*) Use Human and Conspecific Social Cues to Locate Hidden Food," *Journal of Comparative Psychology* 113 (1998): 173–77.

247 **termed "fast mapping":** Juliane Kaminski, Josep Call, and Julia Fischer, "Word Learning in a Domestic Dog: Evidence for 'Fast Mapping,'" *Science* 11 (June 2004): 1682–83, doi:10.1126/science.1097859.

248 **they had probably evolved something he termed "general intelligence":** Charles Darwin, *The Descent of Man* [1871], in *From So Simple a Beginning*, ed. Edward O. Wilson (New York: Norton, 2006), 806.

250 **a ritual to "feed" the soul of the dead animal:** M. Germonpré, M. Lázničková-Galetová, and M. Sablin, "Palaeolithic Dog Skulls at the Gravettian Předmostí Site, the Czech Republic," *Journal of Archaeological Science* 39 (2012): 84–202.

250 **skull of an "incipient dog":** N. D. Ovodov et al., "A 33,000-Year-Old In-

cipient Dog from the Altai Mountains of Siberia: Evidence of the Earliest Domestication Disrupted by the Last Glacial Maximum," *PLoS ONE* 6, no. 7 (2011): e22821, doi:10.1371/journal.pone.0022821.

260 **the child and dog were surely there together . . .:** I first learned about the dog-wolf and child's print from Werner Herzog's brilliant 2010 film, *Cave of Forgotten Dreams.*

EPILOGUE

261 **"What cannot be denied . . .":** Dale Jamieson, "Cognitive Ethology at the End of Neuroscience," in *The Cognitive Animal: Empirical and Theoretical Perspectives on Animal Cognition,* ed. Marc Bekoff, Colin Allen, and Gordon M. Burghardt (Cambridge, MA: MIT Press, 2002), 69–75.

267 **"killjoys," as one prominent philosopher calls them:** Daniel Dennett, quoted in Michael Balter, "'Killjoys' Challenge Claims of Clever Animals," *Science* 335 (2012): 1036, doi: 10.1126/science.335.6072.1036.

267 **"There is grandeur . . .":** Charles Darwin, *On the Origin of Species* [1859], in *From So Simple a Beginning,* ed. Edward O. Wilson (New York: Norton, 2006), 760.

Further Reading

Berg, Karl S., Soraya Delgado, Kathryn A. Cortopassi, Steven R. Beissinger, and Jack W. Bradbury. "Vertical Transmission of Learned Signatures in a Wild Parrot." *Proceedings of the Royal Society B*, July 13, 2011, doi: 10.1098/rspb.2011.0932.

Bradshaw, G. A. *Elephants on the Edge: What Animals Teach Us About Humanity.* New Haven: Yale University Press, 2009.

Braithwaite, Victoria. *Do Fish Feel Pain?* Oxford: Oxford University Press, 2010.

Brown, Culum, Kevin Laland, and Jens Krause. *Fish Cognition and Behavior.* New York: John Wiley and Sons, 2006.

Burghardt, Gordon M. *The Genesis of Animal Play.* Cambridge, MA: MIT Press, 2006.

Connor, Richard C., Rachel A. Smolker, and Andrew Richards. "Two Levels of Alliance Formation Among Male Bottlenose Dolphins (*Tursiops* sp.)." *Proceedings of the National Academy of Sciences* 89 (1992): 987–90.

Crist, Eileen. *Images of Animals: Anthropomorphism and Animal Mind.* Philadephia: Temple University Press, 1999.

Dally, J. M., N. J. Emery, and N. S. Clayton."Food-Caching Western Scrub-Jays Keep Track of Who Was Watching When" *Science* 312 (2006): 1662–65.

Dawkins, Marion Stamp. *Through Our Eyes Only? The Search for Animal Consciousness.* Oxford, UK: Oxford University Press, 1993.

De Waal, Frans. *The Ape and the Sushi Master: Cultural Reflections of a Primatologist.* New York: Basic Books, 2001.

——. "What Is an Animal Emotion?" In *The Year in Cognitive Neuroscience, Annals of the New York Academy of Sciences* 1224 (2011): 191–206.

Foerder, Preston, Marie Galloway, Tony Barthel, Donald E. Moore III, and Diana Reiss. "Insightful Problem Solving in an Asian Elephant." *PLos One* 6 (2011): e23251, doi:10.1371/journal.pone.0023251.

Gácsi, Márta, Borbála Győri, Zsófia Virányi, Enikő Kubinyi, Friederike Range, et al. "Explaining Dog Wolf Differences in Utilizing Pointing Gestures: Selection for Synergistic Shifts in the Development of Some Social Skills." *PLoS One* 4 (2009): e6584, doi:10.1371/journal.pone.0006584.

Garcia, Michel-Alain. "Human Footprints in the Chauvet Cave." *International Newsletter on Rock Art*, no. 24 (1999): 43.

Goodall, Jane. *The Chimpanzees of Gombe: Patterns of Behavior.* Cambridge, MA: Harvard University Press, 1986.

Griffin, Donald R. *Animal Minds: Beyond Cognition to Consciousness.* Chicago: University of Chicago Press, 1992.

Herman, Louis M. "Exploring the Cognitive World of the Bottlenosed Dolphin." In *The Cognitive Animal: Empirical and Theoretical Perspectives on Animal Cognition*, edited by Marc Bekoff, Colin Allen, and Gordon M. Burghardt, Cambridge, MA: MIT Press, 2002, 269–83.

Lonsdorf, Elizabeth V., Stephen R. Ross, and Tetsuro Matsuzawa. *The Mind of the Chimpanzee.* Chicago: University of Chicago Press, 2010.

Mann, Janet, Richard C. Connor, Peter L. Tyack, and Hal Whitehead, eds. *Cetacean Societies: Field Studies of Dolphins and Whales.* Chicago: University of Chicago Press, 2000.

McComb, Karen, Anna Taylor, Christian Wilson, and Benjamin D. Charlton. "Manipulation by Domestic Cats: The Cry Embedded Within the Purr." *Current Biology* 19 (13): R507-R508.

Miklósi, Ádám. *Dog Behaviour, Evolution, and Cognition.* Oxford: Oxford University Press, 2007.

Morell, Virginia. "Dogged." *Smithsonian* (October 2007): 40–42.

———. "Killer Whales Earn Their Name." *Science* 331 (2011): 274–76, doi:10.1126/science.331.6015.274.

Moss, Cynthia J., Harvey Croze, and Phyllis C. Lee. *The Amboseli Elephants: A Long-Term Perspective on a Long-Lived Mammal.* Chicago: University of Chicago Press, 2011.

Panksepp, Jaak. *Affective Neuroscience: The Foundations of Human and Animal Emotions.* Oxford, UK: Oxford University Press, 1998.

Pepperberg, Irene. *The Alex Studies.* Cambridge, MA: Harvard University Press, 1999.

Plotknik, Joshua M., Frans B. M. de Waal, and Diana Reiss. "Self-Recognition in an Asian Elephant." *Proceedings of the National Academy of Sciences* 103 (2006): 17053–57, doi: 10.1073/pnas.0608062103.

Proops, Leanne, Karen McComb, and David Reby. "Cross-Modal Individual Recognition in Domestic Horses." *Proceedings of the National Academy of Sciences*, doi: 106 (2009): 947–51.

Randic, Srdan, Richard C. Connor, William B. Sherwin, and Michael Krutzen. "A Novel Mammalian Social Structure in Indo-Pacific Bottlenose Dolphins (*Tursiops* sp.): Complex Male Alliances in an Open Social Network." *Proceedings of the Royal Society B*, March 28, 2012, doi:10.1098/rspb.2012.0264.

Reiss, Diana. *The Dolphin in the Mirror: Exploring Dolphin Minds and Saving Dolphin Lives.* New York: Houghton Mifflin Harcourt, 2011.

Schlegel, T., and S. Schuster. "Small Circuits for Large Tasks: High-Speed Decision-Making in Archerfish." *Science* 319 (2008): 104–6.

Schuster, S. "Quick Guide: Archerfish." *Current Biology* 17 (2007): R494–95.

Schuster, S., S. Wöhl, M. Griebsch, and I. Klostermeier. "Animal Cognition: How Archer Fish Learn to Down Rapidly Moving Targets." *Current Biology* 16 (2006): 378–83.

Smolker, Rachel. *To Touch a Wild Dolphin: A Journey of Discovery with the Sea's Most Intelligent Creatures.* New York: Anchor Books, 2001.

Weiner, Jonathan. *Time, Love, Memory: A Great Biologist and His Quest for the Origins of Behavior.* New York: Vintage Books, 1999.

Weir, A.A.S., J. Chappell, and A. Kacelnik. "Shaping of Hooks in New Caledonian Crows." *Science* 297 (2002): 981, doi: 10.1126/science.1073433.

Index

About the Author

VIRGINIA MORELL is a correspondent for *Science* and a contributor to *National Geographic*, *Smithsonian*, and *Condé Nast Traveler*, among other publications. She is also the author of *Ancestral Passions*, a *New York Times* Notable Book of the Year, and *Blue Nile*; and she is coauthor with Richard Leakey of *Wildlife Wars*.